6th International Winter Conference on Neurodegeneration

Advisory Board

D. B. Calne
R. Horowski
Y. Mizuno
W. H. Poewe
P. Riederer
M. B. H. Youdim

W. Poewe and G. Ransmayr (eds.)

Advances in Research on Neurodegeneration

Volume 6

Springer-Verlag Wien GmbH

Prof. Dr. W. Poewe
Dr. G. Ransmayr
Universitätsklinik für Neurologie, Innsbruck, Österreich

© 1999 Springer-Verlag Wien
Originally published by Springer-Verlag Wien New York in 1999
Softcover reprint of the hardcover 1st edition 1999

Typesetting: Best-set Typesetter Ltd., Hong Kong
Printing: A. Holzhausens Nfg., A-1070 Wien
Graphic design: Ecke Bonk

Printed on acid-free and chlorine-free bleached paper
SPIN: 10711598

With 21 Figures

ISBN 978-3-211-83261-5   ISBN 978-3-7091-6369-6 (eBook)
DOI 10.1007/978-3-7091-6369-6

Preface

From November 20 to 23, 1997 the 6th International Winter Conference on Neurodegeneration was held in Kitzbühel, Austria, to explore recent advances in the treatment of neurodegenerative and neuroimmunological CNS diseases. Following the tradition of the previous conferences a select group of distinguished basic researchers and clinicians spent two days in the quiet Austrian mountain resort with in depth review and discussion of current topics in the fields of neurodegeneration and neuroimmunology focussing on the interface between these two groups of disorders. The main focus was on how to develop new treatment strategies based on current understandings in the molecular biology of maturation of the nervous system and of basic processes governing neurodegenerative disorders, like Parkinson's disease, multiple system atrophy, Huntington's disease and amyotrophic lateral sclerosis as well as CNS demyelination in multiple sclerosis. Starting from classical concepts of symptomatic dopamine replacement therapy in Parkinson's disease, state-of-the-art lectures covered new evolving therapeutic concepts for neurodegenerative disorders: somatic gene therapy, neural transplantation and neuroprotection through pharmacological treatment as well as through deep brain stimulation.

The emphasis on restoration as key feature of future neurological therapy was also highlighted in a session on mechanisms of demyelination and strategies for myelin repair in neurological autoimmune disease.

The present volume includes updated review articles and original papers covering the proceedings of the 6th International Winter Conference. It is our hope that the information will be of interest to a broad community of clinicians and neuroscientists with special interest in neurodegeneration and neuroimmunology. Finally we would like to thank Schering (Germany) and SmithKline Beecham (Germany and Austria) for their generous support of the meeting.

W. Poewe
Innsbruck, November 1998　　　　　　　　　　　　　　　　**G. Ransmayr**

Contents

Cell differentiation in the embryonic mammalian spinal cord

M. Mayer-Pröschel

Huntsman Cancer Institute, Department of Oncological Science
University of Utah, Salt Lake City, UT, U.S.A.

Summary. The acquisition of cell type specific properties in the spinal cord is a process of a sequential restriction in developmental potential. Multipotent neuroepithelial stem cells (NEP cells) can give rise to all the major cell types in the central nervous system. The generation of these multiple cell types occurs via the generation of intermediate precursor cells, which are restricted in their differentiation potential, but are still able to give rise to more than one cell type. These intermediate precursor cells are different from NEP cells and are different from each other. We have identified neuronal restricted precursor cells (NRP's) which can only generate neurons but no longer glial cells and glial restricted precursor cells (GRP's), which give rise to glial cells but not to neurons. These intermediate precursor cells can be purified and expanded in vitro and might offer a new tool for gene discovery, drug screening and transplantation approaches.

Introduction

Neurodegenerative diseases are often characterized by the loss of specialized cell types which are critical for specific brain functions. For example, oligodendrocytes, the myelin forming cells of the central nervous system, are destroyed in brain lesions of multiple sclerosis patients, while the pathology of Parkinson's disease is predominantly a result of the loss of dopaminergic neurons in the striatum. These deficiencies are compounded by the very limited capacity of the adult brain for regeneration, which poses a serious problem for the reconstruction of affected brain areas and ultimately for any therapeutic approach.

The therapeutic use of precursor cells offers a potentially very promising route to the successful treatment of chronic disorders of the central nervous system (CNS). Strategies that employ cells in therapeutic interventions include the use of cells as delivery vehicles for drugs or genes and the use of suitable cells for replacement cell therapy. The identification of human stem cells and intermediate precursor cells (Moyer et al., 1997; Pincus et al., 1998b; Pincus et al., 1998a; Sah et al., 1997), together with an emerging body of evidence which suggests that there are extensive similarities between human

CNS differentiation and the differentiation processes described in rat and mouse models, have given a new dimension to the therapeutic use of cells in replacement therapies. Diseases like Multiple Sclerosis, epilepsy, Parkinson's disease and stroke have all been targeted in clinical trials with cell replacement therapy and preliminary results are promising (reviewed in Gage, 1998; Martinez-Serrano and Bjorklung, 1997). An important recent result that is also likely to broaden the application of cell transplantation is the demonstration that neurite outgrowth and functional connectivity can be achieved in the adult diseased brain (reviewed in Jones and Redpath, 1998).

Although a variety of multipotent stem cells and lineage restricted precursor cells have been characterized and employed in clinical applications over the past 10 years (Gage, 1998; Gage et al., 1995; Luskin and McDermott, 1994; McConnell, 1995; McKay, 1997; Morrison et al., 1997; Stemple and Mahanthappa, 1997; Svendsen et al., 1997), the understanding of the sequence of events during differentiation in normal CNS development and the multiple signals which regulate the individual differentiation events are still not well understood, thereby limiting the efficacy of these approaches. A better understanding of the differentiation process would allow the controlled manipulation of CNS cells in vitro and ultimately the use of optimal cell populations for transplantation. Until recently, a major obstacle in studying CNS cell differentiation in vitro has been the lack of appropriate culture conditions for these cells. Multipotent neuroepithelial cells have been cultivated as EGF-responsive detached neurospheres which differentiate immediately upon attachment to a surface (Alvarez-Buylla and Lois, 1995; Milward et al., 1997; Morshead et al., 1994; Reynolds et al., 1992; Swendsen et al., 1997; Weiss et al., 1996). Although a large body of data suggests that these EGF dependent cells are multipotent self-renewing stem cells which can generate neurons, astrocytes and oligodendrocytes (Hulspas et al., 1997; Reynolds and Weiss, 1996; Vescovi et al., 1993), clonal analysis and cell counts have been difficult to perform as these cells grow as tightly packed, floating spheres. Another class of stem cells present in the CNS are FGF dependent cells which have not been characterized in as much detail as the EGF-responsive counterpart. Available data suggest that these cells also represent multipotent, self-renewing stem cells able to generate neuron and glial cells (Johe et al., 1996; Kalyani et al., 1997; Palmer et al., 1995; Ray et al., 1995). Although the assumption was made that the differentiated cell types do in fact arise from the multipotent population most likely through the generation of lineage restricted precursor cells, the validity of this assumption has not been directly demonstrated in vitro.

Based on studies of differentiation in other regions of the nervous system, two major developmental models have been proposed for these differentiation events (Fig. 1). The first model is based on the idea that the multiplicity of cell types in the adult spinal cord is a reflection of the initial heterogeneity of the developing neuroepithelium. In this model the neuroepithelium consists of a mosaic of precommited stem cells which give rise to multiple progeny of only one type. A variety of experiments suggest however an alternative model for the generation of cell diversity in the CNS. According to this model,

Model 1

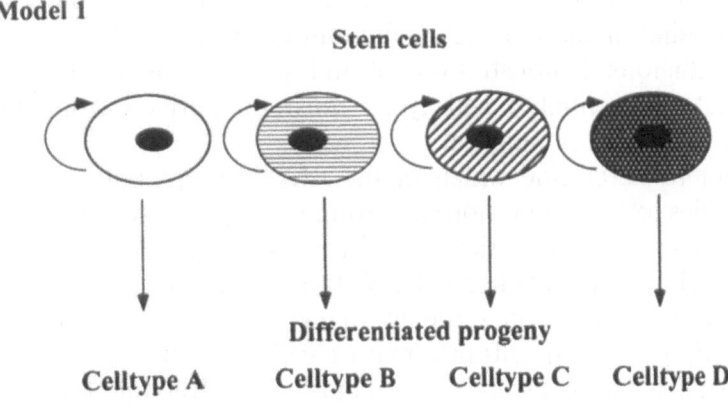

Model 2

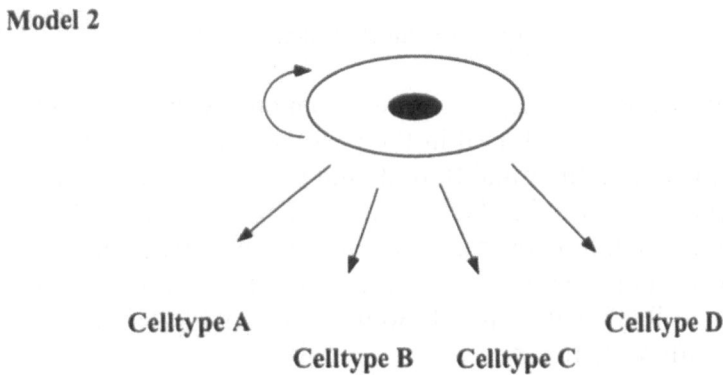

Fig. 1. Stem cell differentiation. Two extreme possibilities for the differentiation of stem cells into mature cell types can be postulated. Model 1 suggests that individual cells may be restricted to generate only a specific subtype of differentiated cells. Their differentiation potential is limited and predetermined. In Model 2 each stem cell is truly multipotential and capable of generating all the major phenotypes of cells present in the brain. The differentiation potential of these stem cells is unlimited and not predetermined. It should be noted that neither model is exclusive. It is conceivable that both types of stem cells exist side-by-side. Circular arrows indicate the ability of cells to self-renew

the neuroepithelium represents a homogenous population of stem cells which all have the ability to generate all the different mature celltypes. Retroviral labeling studies in the developing chick embryo revealed that individual labeled cells can generate multiple phenotypes (Leber et al., 1990; Sanes, 1989). Single injection experiments in chick spinal cords generated labeled progeny that included neuron, glia and PNS derivatives (Artinger et al., 1995; Bronner-Fraser and Fraser, 1988). In addition, Kalyani et al. have recently demonstrated that the majority of the neuroepithelium at embryonic day 10.5 (E10.5) in the rat is also comprised of multipotent stem cells. The authors found that no unipotent cells exist at this stage, though bipotential and tripotential cells are present. This is consistent with findings from other laboratories, who have also demonstrated the existence of multipotent stem cells in the adult spinal cord (Shihabuddin et al., 1995; Weiss et al., 1996). These unambigious demonstrations of multipotent cells in vitro and in vivo suggest that differentiated cell types must be derived from an initially multipotent stem cell population.

In order to determine which of the two developmental differentiation models applies to the generation of mature cell types in the embryonic mammalian spinal cord, we used the cell culture techniques described by Kalyani et al. and FGF-dependent stem cells. With this system we were able to demonstrate and characterize the sequence of events which lead to the differentiation of multipotent stem cells into mature cell types like neurons, astrocytes and oligodendrocytes (Mayer-Proschel et al., 1997; Rao and Mayer-Proschel, 1997; Rao et al., 1997).

The neuroepithelial stem cell

When neuroepithelial cells are isolated from embryonic rat spinal cords (day 10.5 of gestation) and cultured in the presence of chick embryo extract and bFGF they maintain their undifferentiated phenotype for as long as the chick extract is present. Once the chick extract is removed from the cell culture, the cells differentiate into neurons, oligodendrocytes, astrocytes and neural crest cells (Kalyani et al., 1997). This controlled differentiation of multipotent cells into mature cell types in vitro, allowed us to characterize the mechanism by which this differentiation occurs.

We conducted clonal studies which led to the first critical observation that neuroepithelial stem cells represent a homogenous cell population in which each individual cell has the same differentiation potential. None of the stem cells express any known antigenic markers characteristic of differentiated cells. When clones of undifferentiated cells were forced to differentiate, all clones contained both neurons and glial cells (astrocytes and oligodendrocytes) (Rao and Mayer-Proschel, 1997). This observation raised the question whether this differentiation happens as a one step event or whether cells gradually become more and more restricted in their ability to generate a whole range of cell types. The availability of relevant antigenic surface markers made it possible to address this question in vitro.

The generation of glial cells

We observed that the first detectable event after withdrawal of chick embryo extract is the expression of a glial specific ganglioside which is recognized by the monoclonal antibody A2B5 (Dubois et al., 1990; Fok-Seang and Miller, 1994). It is well established that this ganglioside is typically expressed on glial progenitor cells which can give rise to oligodendrocytes and astrocytes depending on environmental cues (Raff et al., 1983). We first tested the possibility that the A2B5[+] population, which is generated from multipotent cells after withdrawal of chick extract represents also a population of glial progenitor cells. The A2B5+ cells were grown in clones and cultured under conditions in which all the relevant celltypes (astrocytes, oligodendrocytes, neurons and neural crest cells) are able to survive. It became clear, that although the medium would support any of the above mentioned cell types, the A2B5+ cells were only able to differentiate into glial cells, i.e. oligodendrocytes or astrocytes. This result suggested very strongly that the glial cells in the early spinal cord are indeed generated through a lineage restricted precursor cell which originates from multipotent neuroepithelial cells. In other words, the mechanism by which postmitotic differentiated cells are generated in the embryonic spinal cord is a stepwise lineage restriction. The A2B5+ cells represent therefore the glial restricted precursor cells (GRP cells) which are no longer able to generate any other cell type but glia. We could subsequently show that this glial restricted precursor cells can be purified by immunopanning techniques and the pure population can be expanded in vitro for at least 3 months. Even after this time period in culture in the presence of appropriate mitogens, the cells do not loose their ability to differentiate into oligodendrocytes or astrocytes (Rao and Mayer-Proschel, 1997; Rao et al., 1997).

In addition to the direct demonstration that multipotent stem cells can give rise to lineage restricted intermediate precursor cells this GRP population might also represent a useful cell population in enabling the repair of lesions using transplantation techniques. Two characteristics make this cell population particularly attractive for the use in transplantation. First, GRP cells can be expanded as pure populations for a very long time in culture thus providing sufficient quantities of cells and secondly, these cells retain the ability to differentiate into mature cell types even after prolonged in vitro expansion.

The generation of neurons

In addition to pathological conditions predominantly disrupting glial cells there are a considerable number of pathological conditions in which specific neuron populations are affected. It is therefore of major interest to identify a neuronal counterpart to the glial restricted precursor cell population. We observed that although 70% of all cells express A2B5 after withdrawal of the chick embryo extract the vast majority of the remaining 30% expressed embryonic NCAM (E-NCAM), an antigen which is expressed in early neurogenesis (Blass-Kampmann et al., 1994; Chen and Chiu, 1992). We speculated that this population might represent the neuronal precursor cells which are destined to differentiate into mature neurons. We again performed the clonal analysis of individual E-NCAM+ cells and investigated their differentiation potential under a variety of culture conditions. Both mass culture and clonal experiments showed that the E-CAM+ cells which are generated from the multipotent NEP population represent neuronal precursor cells. In addition, these cells are not only able to generate a variety of neuronal cell types

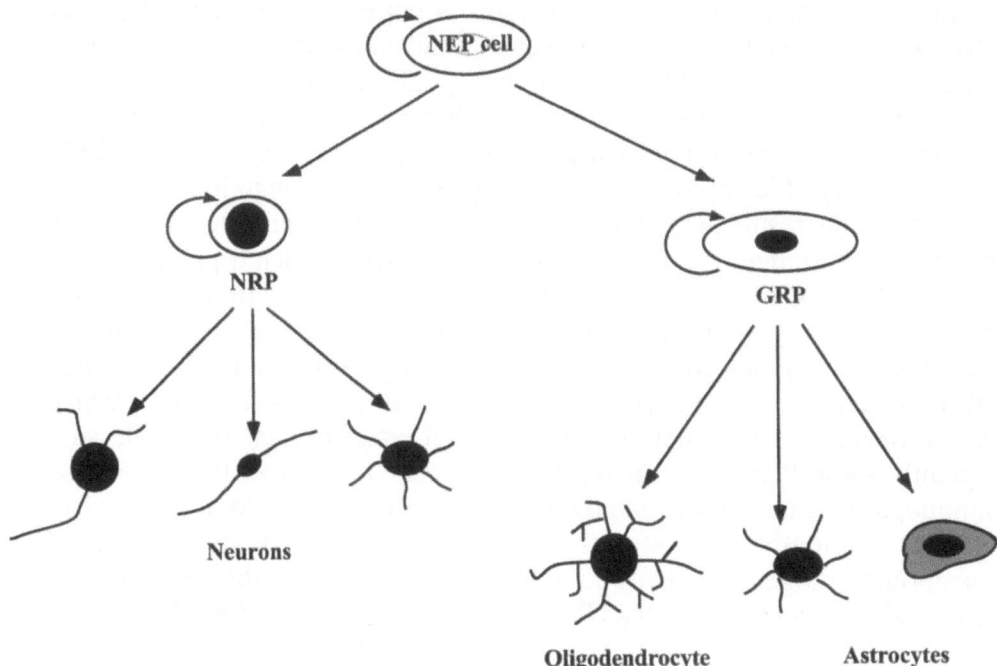

Fig. 2. Relationship of NEP stem cells and intermediate precursor cells. Upon differentiation, NEP cells give rise to neuron restricted precursor cells (*NRP*) and glial restricted precursor cells (*GRP*). These cell populations are all able to extensive selfrenewal (indicated by circular arrow). NRP can generate different phenotypes of neurons and GRP cells can generate oligodendrocytes of two distinct types of astrocytes depending on the environmental signals

but are restricted to the neuronal lineage. We were not able to generate any glial cells from E-NCAM+ cells in any culture condition we tested (Mayer-Proschel et al., 1997).

In summary we have shown that multipotent neuroepithelial cells derived from the rat E10.5 spinal cord are a homogenous cell population in which each individual cell is able to give rise to cells of the neuronal and glial lineage. Differentiated cell types arise from these NEP cells through stepwise differentiation events during which intermediate lineage restricted precursor cells are generated. These fate-restricted cells are able to proliferate for extended periods of time and can give rise to either neuronal phenotypes or to astrocytes and oligodendrocytes, which belong to the glial lineage (Fig. 2). These experiments show that the development of mature celltypes from a multipotent cell pool seems to be governed by a stepwise lineage restriction. The cell culture system we described above now allows the study of the molecular mechanism(s) and signals by which this stepwise restriction is achieved. Extending from these fundamental observations, defined transplantation experiments will determine the usefulness of these lineage restricted cell populations in transplantation approaches.

References

Alvarez-Buylla A, Lois C (1995) Neuronal stem cells in the brain of adult vertebrates. Stem Cells 13: 263–272

Artinger KB, Fraser S, Bronner-Fraser M (1995) Dorsal and ventral types can arise from common neural tube progenitors. Dev Biol 172: 591–601

Blass-Kampmann S, Reinhardt-Maelicke S, Kindler-Rohrborn A, Cleeves V, Rajewsky MF (1994) In vitro differentiation of E-NCAM expressing rat neural precursor cells isolated by FACS during prenatal development. J Neurosci Res 37: 359–373

Bronner-Fraser M, Fraser SE (1988) Cell lineage analysis shows multipotentiality of some avain neural crest cells. Nature 355: 161–164

Chen EW, Chiu AY (1992) Early stages in the development of spinal motor neurons. Comp Neurosci 320: 291–303

Dubois C, Manuguerra JC, Hauttecoeur B, Maze J (1990) Monoclonal antibody A2B5, which detects cell surface antigens, binds to ganglioside GT3 (II3 (NeuAc)3LacCer) and to its 9-O-acetylated derivative. J Biol Chem 265: 2797–2803

Fok-Seang J, Miller RH (1994) Distribution and differentiation of A2B5+ glial precursors in the developing rat spinal cord. J Neurosci Res 37: 219–235

Gage FH (1998) Cell therapy. Nature 392: 18–24

Gage FH, Ray J, Fisher LJ (1995) Isolation, characterisation and use of stem cells from the CNS. Annu Rev Neurosci 18: 159–192

Hulspas R, Tiarks C, Reilly J, Hsieh CC, Recht L, Quesenberry PJ (1997) In vitro cell density-dependent clonal grwoth of EGF-responsive murine neural progenitor cells under scrum-frcc conditions. Exp Ncurol 148: 147–156

Johe KK, Hazel TG, Muller T, Dugich Djordjevic MM, McKay RD (1996) Single factors direct the differentiation of stem cells from the fetal and adult central nervous system. Genes Dev 10: 3129–3140

Jones DG, Redpath CM (1998) Regeneration in the central nervous system: pharmacological intervention, xenotransplantation and stem cell transplantation. Clin Anat 11: 263–270

Kalyani A, Hobson K, Rao MS (1997) Neuroepithelial stem cells from the embryonic spinal cord: isolation, characterization and clonal analysis. Dev Biol 187: 203–226

Leber SM, Breedlove SM, Sanes JR (1990) Lineage, arrangement, and death of clonally related motoneurons in chick spinal cord. J Neurosci 10: 2451–2462

Luskin MB, McDermott K (1994) Divergent lineages for oligodendrocytes and astrocytes originating in the neonatal forebrain subventricular zone. Glia 11: 211–226

Martinez-Serrano A, Bjorklung A (1997) Immortalized neural progenitor cells for CNS gene transfer and repair. TINS 20: 530–537

Mayer-Proschel M, Kalyani AJ, Mujtaba T, Rao M (1997) Isolation of lineage-restricted neuronal precursors from multipotent neuroepithelial stem cells. Neuron 19: 773–785

McConnell SK (1995) Strategies for the generation of neuronal diversity in the developing central nervous system. J Neurosci 15: 6987–6998

McKay R (1997) Stem cells in the central nervous system. Science 276: 66–71

Milward EA, Lundberg CG, Ge B, Lipsitz D, Zhao M, Duncan ID (1997) Isolation and transplantation of multipotential populations of epidermal growth factor-responsive, neural progenitor cells from the canine brain. J Neurosci Res 50: 862–871

Morrison SJ, Shah NM, Anderson DJ (1997) Regulatory mechanism in stem cell biology. Cell 88: 287–298

Morshead CM, Reynolds BA, Craig CG, McBurney MW, Staines WA, Morassutti D, Weiss S, van der Kooy D (1994) Neural stem cells in the adult mammalian forebrain: a relatively quiescent subpopulation of the subependymal cells. Neuron 13: 1071–1082

Moyer MP, Johnson RA, Zompa EA, Cain L, Morshed T, Hulsebosch CE (1997) Culture, expansion and transplantation of human fetal neural progenitor cells. Transplant Proc 29: 2040–2041

Palmer TD, Ray J, Gage F (1995) FGF-responsive neuronal progenitors reside in proliferative and quiescent regions of the adult rodent brain. Mol Cell Neurosci 6: 474–486

Pincus DW, Keyoung HM, Harrison-Restelli C, Goodman RR, Fraser RA, Edgar M,
Sakakibara S, Okano H, Nedergaard M, Goldman SA (1998a) Fibroblast Growth
Factor-2/brain derived neurotrophic factor assicaited maturation of new neurons
generated from adult human suependymal cells. Ann Neurol 43: 576–585

Pincus DW, Goodman RR, Fraser RA, Nedergaard M, Goldman SA (1998b) Neural
stem and progenitor cells: a strategy for gene therapy and brain repair. Neurosurgery
42: 858–867

Raff MC, Miller RH, Noble M (1983) A glial progenitor cell that develops in vitro into an
astrocyte or an oligodendrocyte depending on the culture medium. Nature 303: 390–
396

Rao M, Mayer-Proschel M (1997) Glial restricted precursors are derived from multipotent neuroepithelial stem cells. Dev Biol 188: 48–63

Rao M, Noble M, Mayer-Proschel M (1998) A tripotential glial precursor cell is present
in the developing spinal cord. PNAS 95: 3996–4001

Ray J, Raymond HK, Gage FH (1995) Generation and culturing of precursor cells and
neuroblasts from embryonic and adult central nervous system. Methods Enzymol
254: 20–37

Reynolds BA, Weiss S (1996) Clonal and population analyses demonstrate that an EGF-
responsive mammalian embryonic CNS precursor is a stem cell. Dev Biol 175: 1–13

Reynolds BA, Tetzlaff W, Weiss S (1992) A mulitpotent EGF-responsive striatal embryonic progenitor cell produces neurons and astrocytes. J Neurosci 12: 4565–4574

Sah DW, Ray J, Gage FH (1997) Bipotent progenitor cell lines from the human CNS. Nat
Biotechnol 15: 574–580

Sanes JR (1989) Analysing cell lineages with a recombinant retrovirus. TINS 12: 21–28

Shihabuddin LS, Hertz JA, Holets VR, Whittemore SR (1995) The adult CNS retains the
potential to direct region-specific differentiation of a transplanted neuronal precursor
cell line. J Neurosci 15: 6666–6678

Stemple DL, Mahanthappa NK (1997) Neural stem cells are blasting off. Neuron 18: 1–4

Svendsen CN, Caldwell MA, Shen J, ter Borg MG, Rosser AE, Tyers P, Karmiol S,
Dunnett SB (1997) Long-term survival of human central nervous system progenitor
cells transplanted into a rat model of Parkinson's disease. Exp Neurol 148: 135–146

Swendsen CN, Clarke DJ, Rosser AE, Dunnett SB (1997) Survival and differentiation of
rat and human epidermal growth factor-responsive precursor cells following grafting
into the lesioned adult central nervous system. Exp Neurol 137: 376–388

Vescovi AL, Reynolds BA, Fraser DD, Weiss S (1993) bFGF regulates the proliferative
fate of unipotent (neuronal) and bipotent (neuron/astroglial) EGF-generated CNS
progenitor cells. Neuron 11: 951–966

Weiss S, Dunne C, Hewson J, Wohl C, Wheatley M, Peterson AC, Reynolds BA (1996)
Multipotent CNS stem cells are present in the adult mammalian spinal cord and
ventricular neuroaxis. J Neurosci 16: 7599–7609

Author's address: Margot Mayer-Pröschel, Huntsman Cancer Institute at the
University of Utah, Biomedical Polymers Research Building, Room 410C, Salt Lake
City, UT 84112, U.S.A.

Myelin dysfunction/degradation in the central nervous system: why are myelin sheaths susceptible to damage?

M. Bradl

Max-Planck-Institute for Neurobiology, Department of Neuroimmunology,
Martinsried, Federal Republic of Germany

Summary. In the central nervous system, myelin sheaths are produced to electrically insulate axons and to increase the velocity of axonal conduction. They are highly complex structures, which are often destructed in neurological disorders. One possible reason for the vulnerability of myelin sheaths to damage became apparent from analyses of animals with altered amounts of otherwise normal myelin components: Due to limited redundance in function between different myelin proteins, dysfunction or loss of one protein may cause loss of function and instability of the entire myelin sheath.

Introduction

Destruction of myelin sheaths in the central nervous system (CNS) is a common theme in many neurological disorders, with a wide range of underlying causes. For example, demyelination can result from CNS inflammation, where it is caused by T cell/macrophage infiltration, especially when this is associated with myelin-specific autoantibody responses (Linington et al., 1988), or by pro-inflammatory mediators such as cytokines (Probert et al., 1995; Corbin et al., 1996), complement (Dietzschold et al., 1995) and free radicals (Brett and Rumsby, 1993; Bagasra et al., 1995). Alternatively, myelin destruction can be triggered by viral infections (Rodriguez, 1985; Rodriguez et al., 1996; Fleming et al., 1993). Finally, it can be caused by defects of genes encoding structural or regulatory myelin proteins (Duncan, 1995). But why are myelin sheaths so extremely vulnerable?

In the CNS, myelin sheaths are produced by oligodendrocytes to electrically insulate axons and thus increase the velocity of axonal conduction. They are comprised of multiple layers of oligodendrocytic plasma membrane, and contain many proteins highly specific to CNS myelin (Fig. 1).

The important contribution of individual myelin components to myelin stability and function is immediately evident from a broad spectrum of natural mutations resulting in unstable myelin and corresponding behavioral abnormalities (for review see Duncan, 1995). However, usage of such mutants to deduce the function of individual myelin components is limited by one impor-

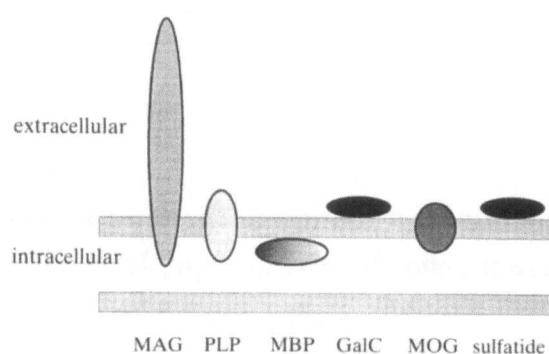

Fig. 1. Myelin components in the mammalian CNS

tant draw-back: mutated proteins could either perform new functions (gain-of-function phenotypes), could be unable to perform some of their functions (loss-of-function phenotypes) or could interfere with the proper performance of other components of the myelin sheath.

One way to circumvent such problems is to analyze animals with altered amounts of otherwise normal myelin components. Occasionally, such animals are natural mutants (for example mice carrying the mutation *shiverer*, see below). More often, however, they had been genetically manipulated — either by increasing the gene dosage with introduced transgenes, or by targeted disruption ("knock-out") of relevant gene loci.

What did we learn from such animals about the function of individual myelin constituents?

Myelin basic protein (MBP)

Reduction, or even complete absence of MBP is seen in mice carrying the autosomal recessive mutations shiverer (shi) or shiverer myelin deficient (shi[mld]), respectively. In shiverer mice, a large portion of the MBP gene is deleted, leading to a complete absence of functional MBP products (Roach et al., 1983; Roach et al., 1985; Kimura et al., 1985). In shi[mld], the entire MBP gene is duplicated, and some of its exons are inverted, resulting in antisense RNA for MBP transcripts and, eventually, in a significant reduction of total MBP production (Popko et al., 1987, 1988). MBP deficient animals can be easily recognized by a characteristic "shivering" which is most evident when they initiate voluntary movements. The CNS of these mutants is characterized by apparently normally differentiated oligodendrocytes, and a profound reduction in the amount of myelin. Only few, disproportionally thin myelin sheaths with an atypical ultrastructure can be detected. These sheaths lack a major dense line — the intracellular adhesion zone of the myelin membrane (Roach et al., 1983). Shi and shi[mld] are natural mutants of the mouse fancy. Their phenotype was faithfully reproduced in transgenic animals upon MBP reduction by antisense technology (Katsuki et al., 1988), and corrected

Table 1. Function of proteins and lipids in CNS myelin, as revealed by transgenic overexpression or genetic ablation

Component	Alteration	Principal observation	Conclusion
CNP Gravel et al. (1996)	overex	temporally accelerated MBP and PLP expression lack of myelin compaction	CNP regulator of oligodendrocyte maturation
GalC and sulfatides Bosio et al. (1996) Coetzee et al. (1996) Dupree et al. (1998)	cgt-k.o.	myelin compact, but instable, insulator function of myelin sheaths lost, abnormal nodes (Dupree et al., 1998)	GalC/sulfatides important for myelin stability and node of Ranvier formation
MAG Montag et al. (1994) Li et al. (1994)	MAG-k.o.	delayed onset of myelination; some axons with two or more sheaths (Montag et al. 1994) reduced content of oligo-cyte cytoplasm at inner aspects of most myelin sheaths (Montag et al., 1994; Li et al., 1994)	L-MAG necessary for glial-axonal interactions; participation in recognition between oligodendrocyte processes and axons
MBP Katsuki et al. (1988) Shine et al. (1992)	MBP-red shi shi[mld]	Reduction in CNS myelin presence of well differentiated oligodendrocytes myelin sheaths lack major dense line (Shine et al., 1992)	MBP necessary for compaction of myelin
PLP Boison and Stoffel (1994) Boison et al. (1995)	k.o.	reduced myelin stability	PLP needed for stabilization of compacted myelin
PLP Readhead et al. (1994) Kagawa et al. (1994)	overex	overexpression by factor >2: oligodendrocytes apoptotic and developmentally arrested profound lack of myelin sheaths Overexpression by factor <2: myelin degeneration (Kagawa et al., 1994)	PLP regulator of oligodendrocyte differentiation/survival; needed for maintaining myelin sheaths
DM-20 Mastronardi et al. (1993) Simons-Johnson et al. (1995)	overex	reduced myelin content in CNS, disrupted myelin lamellae late onset demyelination with lymphocytic infiltrates	DM-20 needed for maintaining myelin sheaths

upon introduction of wildtype MBP sequences (Readhead et al., 1987; Kimura et al., 1989), indicating first, that the amount of MBP available to oligodendrocytes is a limiting factor for CNS myelin assembly, and secondly, that the presence of MBP is crucial for the compaction of myelin (Table 1).

Since lack of MBP had such profound consequences for myelination in the CNS, it was interesting to see, whether complete absence of proteolipid protein (PLP) would also be deleterious to assembly or maintenance of intact myelin sheaths.

Proteolipid protein (PLP)

It was known for a long time that the mouse mutation jimpy (jp) and the corresponding rat mutation myelin deficient (md) affect the PLP gene, leading to abnormally folded and rapidly degraded PLP (Nave et al., 1986; Boison and Stoffel, 1989). Animals carrying these mutations display abnormal numbers of apoptotic oligodendrocytes, associated with activated microglia cells and astrocytes. The CNS of these mutants is severely dysmyelinated, and the few remaining myelin sheaths do not contain immunoreactive PLP (for review see Nave, 1995). Since jp mice and md rats had such severe changes in their phenotype, it was quite surprising to see that oligodendrocytes from PLP knock-out mice were viable and able to ensheath axons of all calibers with compacted myelin (Klugmann et al., 1997; Boison and Stoffel, 1994), and that PLP deficiency even rescued the lethal phenotype of shiverer mutants (Stoffel et al., 1997). It needed careful analyses by electron microscopy to reveal a lack of firm intermembrane bonding and a tendency of CNS PLP$^{-/-}$ myelin to losen up (Rosenbluth et al., 1996). These experiments clearly demonstrated that PLP is needed to ensure proper physical stability of myelin sheaths. But there was more to be learned about the function of PLP in the CNS. Transgenic mice (Readhead et al., 1994; Kagawa et al., 1994) and rats (own, unpublished results) overexpressing PLP were produced. These animals have two different phenotypes, depending on the degree of overexpression.

Expressing proteolipids above a certain threshold (i.e. expressing at least double the amount than normal) resulted in the death of mature, and a developmental arrest of remaining immature oligodendrocytes. Consequently, the CNS of these animals is severely dysmyelinated. Death of mature oligodendrocytes is almost certainly consequence of retention and accumulation of incorrectly folded PLP proteins in the endoplasmic reticulum — a condition also described in mice and rats carrying PLP missense mutations (Gow et al., 1994). However, the developmental arrest of immature, premyelinating oligodendrocytes strongly suggests that PLP has an important role in cellular differentiation as well, possibly by forming ion channels. This speculation is supported by a high similarity of proteolipids to pore-forming proteins (Yan et al., 1993; Kitagawa et al., 1993), although formal proof for it is lacking to date.

Important information was also gained from mice and rats with lower levels of PLP overexpression (i.e. overexpressing PLP by factor <2). At first glance, these animals seemed entirely healthy and were fully competent to produce normal myelin sheaths. However, these sheaths were extremely susceptible to degeneration and/or frank demyelination (Readhead et al., 1994; Kagawa et al., 1994, own unpublished results), indicating that a correct stoichiometric ratio of PLP in the myelin membrane is crucial for maintaining myelin sheaths.

Myelin-associated glycoprotein (MAG)

PLP and MBP are major myelin proteins, and so it is just natural that they had been analyzed first. Other proteins were to follow soon, though. Mice were produced with a targeted deletion of the MAG gene. These animals were viable (Li et al., 1994; Montag et al., 1994), but showed a delay in CNS myelin formation (Montag et al., 1994). Moreover, their CNS myelin was abnormal: cytoplasmic collars of oligodendrocytes usually present in mature sheaths were either completely missing, or significantly reduced in numbers, and many axons were surrounded by 2 or more sheaths. Subsequently it could be shown, that MAG is also an inhibitory compontent for axonal growth in peripheral (Schäfer et al., 1996), but not central myelin (Bartsch et al., 1995). These data indicated that MAG plays a critical role in glial-axonal interactions. But there were two more points to consider. First, aged MAG[-/-] mutants displayed evidence of dying-back oligodendrogliopathy (Lassmann et al., 1997) in the CNS, and of myelin degeneration in the PNS (Fruttiger et al., 1995). Secondly, MAG exists in 2 distinct isoforms, as larger L-MAG and as the shorter splice variant S-MAG. To analyze the contribution of individual MAG isoforms to CNS and PNS pathology, a novel set of "knock-out" mice had to be produced, which carried dysfunctional L-MAG, and normal S-MAG. L-MAG[-/-] were viable, and displayed most of the CNS, but none of the PNS abnormalities (Fujita et al., 1998). Thus, L-MAG is responsible for the integrity of myelin sheaths in the CNS. Interestingly, a profund decrease in L-MAG and myelin malformations (Fujita et al., 1990; Bö et al., 1995) is also seen in mice carrying the mutation quaking (qk), which affects a protein possibly involved in RNA processing and signal transduction (Ebersole et al., 1996).

Galactocerebroside (Gal-C) and sulfatides

All myelin components described so far were proteins. Also lipid components of the myelin sheath could be manipulated, by indirect means. This was achieved by genetically disrupting UDP-galactose ceramide galactosyltransferase (cgt), a key enzyme for the synthesis of Gal-C and sulfatides. Accordingly, these glycoplipids are completely absent in cgt[-/-] mice (Coetzee et al., 1996; Bosio et al., 1996). Although affected animals exhibit a severe tremor that is associated with hind limb paralysis, most of their myelin sheaths are remarkably intact. Nevertheless, axonal conduction velocity was significantly decreased, due to a defect in the formation of nodes of Ranvier in the CNS (Dupree et al., 1998). Thus, Gal-C and sulfatides are important in ensuring proper glial-axonal interactions.

Why are myelin sheaths vulnerable?

As pointed out above, myelin sheaths are highly complex, almost crystalline arrangements of lipids, glycolipids and proteins. The balance of these individual elements seems so delicately tuned, that not only overt mutations, but even slightest increases or decreases in the amount of otherwise normal components perturb the entire structure (Scherer, 1997). Contrary to other systems, where dysfunction or loss of one protein could be compensated for by the action of others (Erickson, 1993; Kelso, 1994), myelin components do not seem to have such a "back-up". Interfering with the function of one indivual component — for whatever reasons — may cause a loss of function of the whole structure, making myelin sheaths to prime targets in many different neurological disorders.

References

Bagasra O, Michaels FH, Zheng YM, Bobroski LE, Spitsin SV, Fu ZF, Tawadros R, Koprowski H (1995) Activation of the inducible form of nitric oxide synthase in the brains of patients with multiple sclerosis. Proc Natl Acad Sci USA 92: 12041–12045

Bartsch U, Bandtlow CE, Schnell L, Bartsch S, Spillmann AA, Rubin BP, Hillenbrand R, Montag D, Schwab ME, and Schachner M (1995) Lack of evidence that the myelin-associated glycoprotein (MAG) is a major inhibitor of axonal regeneration in the CNS. Neuron 15: 1375–1382

Boison D, Stoffel W (1989) Myelin deficient rat: A point mutation in exon III (A > C, Thr75 > Pro) of the myelin proteolipid protein causes dysmyelination and oligodendrocyte death. EMBO J 8: 3295–3302

Boison D, Stoffel W (1994) Disruption of the compacted myelin sheath of axons of the central nervous system in proteolipid protein-deficient mice. Proc Natl Acad Sci USA 91: 11709–11713

Boison D, Büssow H, D'Urso D, Müller HW, Stoffel W (1995) Adhesive properties of proteolipid protein are responsible for the compaction of CNS myelin sheaths. J Neurosci 15: 5502–5513

Bosio A, Binczek E, Stoffel W (1996) Functional breakdown of the lipid bilayer of the myelin membrane in central and peripheral nervous system by disrupted galactocerebroside synthesis. Proc Natl Acad Sci USA 93: 13280–13285

Bö L, Quarles RH, Fujita N, Bartoszewicz Z, Sato S, Trapp BD (1995) Endocytic depletion of L-MAG from CNS myelin in quaking mice. J Cell Biol 131: 1811–1820

Brett R, Rumsby MG (1993) Evidence of free radical damage in the central nervous system of guinea pigs at the prolonged acute and early relapse stages of chronic relapsing experimental allergic encephalomyelitis. Neurochem Int 23: 35–44

Coetzee T, Fujita N, Dupree J, Shi R, Blight A, Szuzuki K, Suzuki K, Popko B (1996) Myelination in the absence of galactocerebroside and sulfatide: Normal structure with abnormal function and regional instability. Cell 86: 209–219

Corbin JG, Kelly D, Rath EM, Baerwald KD, Suzuki K, Popko B (1996) Targeted CNS expression of interferon-y in transgenic mice leads to hypomyelination, reactive gliosis, and abnormal cerebellar development. Mol Cell Neurosci 7: 354–370

Dietzschold B, Schwaeble W, Schäfer MK-H, Hooper DC, Zehng YM, Petry F, Sheng H, Fink T, Loos M, Koprowski H, Weihe E (1995) Expression of C1q, a subcomponent of the rat complement system, is dramatically enhanced in brains with either Borna disease or experimental allergic encephalomyelitis. J Neurol Sci 130: 11–16

Duncan ID (1995) Inherited disorders of myelination of the central nervous system. In: Kettenmann H, Ransom B (eds) Neuroglia. Oxford University Press, Oxford, U.K., pp. 990–1009

Dupree JL, Coetzee T, Blight A, Suzuki K, Popko B (1998) Myelin galactolipids are essential for proper node of Ranvier formation in the CNS. J Neurosci 18: 1642–1649

Ebersole TA, Chen Q, Justice MJ, Artzt K (1996) The quaking gene product necessary in embryogenesis and myelination combines features of RNA binding and signal transduction proteins. Nat Genet 12: 260–265

Erickson HP (1993) Gene knockouts of c-src, transforming growth factor b1, and tenascin suggest superfluous, nonfunctional expression of proteins. J Cell Biol 120: 1079–1081

Fleming JO, Wang FI, Trousdale MD, Hinton DR, Stohlman SA (1993) Interaction of immune and central nervous systems: Contribution of anti-viral Thy-1[+] cells to demyelination induced by coronavirus JHM. Reg Immunol 5: 37–43

Fruttiger M, Montag D, Schachner M, Martini R (1995) Crucial role for the myelin-associated glycoprotein in the maintenance of axon-myelin integrity. Eur J Neurosci 7: 511–515

Fujita N, Sato S, Ishigura H, Inuzuka T, Baba H, Kurihara T, Takahashi Y, Miyatake T (1990) The large isoform of myelin-associated glycoprotein is scarcely expressed in the quaking mouse brain. J Neurochem 55: 1056–1059

Fujita N, Kemper A, Dupree J, Nakayasu H, Bartsch U, Schachner M, Maeda N, Suzuki K, Suzuki K, Popko B (1998) The cytoplasmic domain of the large myelin-associated glycoprotein isoform is needed for proper CNS but not peripheral nervous system myelination. J Neurosci 18: 1970–1978

Gow A, Friedrich VL, Lazzarini RA (1994) Many naturally occurring mutations of myelin proteolipid protein impair its intracellular transport. J Neurosci Res 37: 574–583

Gravel M, Peterson J, Yong VW, Kottis V, Trapp BD, Braun PE (1996) Overexpression of 2′,3′-cyclic nucleotide 3′-phosphodiesterase in transgenic mice alters oligodendrocyte development and produces aberrant myelination. Mol Cell Neurosci 7: 453–466

Johnson RS, Roder JC, Riordan JR (1995) Over-expression of the DM-20 myelin proteolipid causes central nervous system demyelination in transgenic mice. J Neurochem 64: 967–976

Kagawa T, Ikenaka K, Inoue Y, Kuriyama S, Tsujii T, Nakao J, Nakajima K, Aruga J, Okano H, Mikoshiba K (1994) Glial cell degeneration and hypomyelination caused by overexpression of myelin proteolipid protein gene. Neuron 13: 427–442

Katsuki M, Sato M, Kimura M, Yokoyama M, Kobayashi K, Nomura T (1988) Conversion of normal behavior to shiverer by myelin basic protein antisense cDNA in transgenic mice. Science 241: 593–393

Kelso A (1994) The enigma of cytokine redundancy. Immunol Cell Biol 72: 97–101

Kimura M, Inoko H, Katsuki M, Ando A, Sato T, Hirose T, Takashima H, Inayama S, Okano H, Takamatsu K, Mikoshiba K, Tsukuda Y, Watanabe I (1985) Molecular genetic analysis of myelin deficient mice: Shiverer mutant mice show deletion in gene(s) coding for myelin basic protein (MBP). J Neurochem 44: 692–696

Kimura M, Sato M, Akatsuka A, Nozawa-Kimura S, Takahashi R, Yokoyama N, Nomura T, Katsuki M (1989) Restoration of myelin formation by a single type of myelin basic protein (MBP) in transgenic shiverer mice. Proc Natl Acad Sci USA 86: 5661–5665

Kitagawa K, Sinoway MP, Yang C, Gould RM, Colman DR (1993) A proteolipid protein gene family: Expression in sharks and rays and possible evolution from an ancestral gene encoding a pore-forming polympeptide. Neuron 11: 433–448

Klugmann M, Schwab MH, Pühlhofer A, Schneider A, Zimmermann F, Griffiths IR, Nave K-A (1997) Assembly of CNS myelin in the absence of proteolipid protein. Neuron 18: 59–70

Lassmann H, Bartsch U, Montag D, Schachner M (1997) Dying back oligoden-drogliopathy: A late sequel of myelin-associated glycoprotein deficiency. Glia 19: 104–110

Li C, Tropak MB, Gerlai R, Clapoff S, Abramow-Newerly W, Trapp BD, Peterson A, Roder J (1994) Myelination in the absence of myelin-associated glycoprotein. Nature 369: 747–750

Linington C, Bradl M, Lassmann H, Brunner C, Vass K (1988) Augmentation of demy-elination in rat acute allergic encephalomyelitis by circulating mouse monoclonal antibodies directed against a myelin/oligodendrocyte glycoprotein. Am J Pathol 130: 443–454

Mastronardi FG, Ackerley CA, Arsenault L, Roots BI, Moscarello MA (1993) Demyeli-nation in a transgenic mouse: A model for multiple sclerosis. J Neurosci Res 36: 315–324

Montag D, Giese KP, Bartsch U, Martini R, Lang Y, Bluethmann H, Karthigasan J, Kirschner DA, Wintergerst E, Nave K-A, Zielasek J, Toyka KV, Lipp H-P, Schachner M (1994) Mice deficient for the myelin-associated glycoprotein show subtle abnormalities in myelin. Neuron 13: 229–246

Nave K-A (1995) Neurological mouse mutants: a molecular-genetic analysis of myelin proteins. In: Kettenmann H, Ransom B (eds) Neuroglia. Oxford University Press, Oxford, U.K., pp. 571–586

Nave K-A, Lai C, Bloom FE, Milner RJ (1986) Jimpy mutant mouse: A 74-base deletion in the mRNA for myelin proteolipid protein (PLP) and evidence for a primary defect in RNA splicing. Proc Natl Acad Sci USA 83: 9264–9268

Popko B, Puckett C, Lai E, Shine HD, Readhead C, Takahashi N, Hunt III SW, Sidman R, Hood LE (1987) Myelin deficient mice: Expression of myelin basic protein and generation of mice with varying levels of myelin. Cell 48: 713–721

Popko B, Puckett C, Hood LE (1988) A novel mutation in myelin-deficient mice results in unstable myelin basic protein gene transcripts. Neuron 1: 221–225

Probert L, Akassoglou K, Pasparakis M, Kontogeorgos G, Kollias G (1995) Spontaneous inflammatory demyelinating disease in transgenic mice showing central nervous system-specific expression of tumor necrosis factor a. Proc Natl Acad Sci USA 92: 11294–11298

Readhead C, Popko B, Takahashi N, Shine HD, Saavedra RA, Sidman RL, Hood L (1987) Expression of a myelin basic protein gene in transgenic shiverer mice: correc-tion of the dysmyelinating phenotype. Cell 48: 703–712

Readhead C, Schneider A, Griffiths I, Nave K-A (1994) Premature arrest of myelin formation in transgenic mice with increased proteolipid protein gene dosage. Neuron 12: 583–595

Roach A, Boylan K, Horvath S, Prusiner SB, Hood LE (1983) Characterization of cloned cDNA representing rat myelin basic protein: Absence of expression in brain of shiverer mutant mice. Cell 34: 799–806

Roach A, Takahashi N, Pravtcheva D, Ruddle F, Hood L (1985) Chromosomal mapping of mouse myelin basic protein gene and structure and transcription of the partially deleted gene in shiverer mutant mice. Cell 42: 149–155

Rodriguez M (1985) Virus-induced demyelination in mice: "Dying-back" of oligodendro-cytes. Mayo Clin Proc 60: 433–438

Rodriguez M, Pavelko KD, Njenga MK, Logan WC, Wettstein PJ (1996) The balance between persistent virus infection and immune cells determines demyelination. J Immunol 157: 5699–5709

Rosenbluth J, Stoffel W, Schiff R (1996) Myelin structure in proteolipid protein (PLP)-null mouse spinal cord. J Comp Neurol 371: 336–344

Schäfer M, Fruttiger M, Montag D, Schachner M, Martini R (1996) Disruption of the gene for the myelin-associated glycoprotein improves axonal regrowth along myelin in C57BL/Wld^s mice. Neuron 16: 1107–1113

Scherer S (1997) Molecular genetics of demyelination: New wrinkles on an old membrane. Neuron 18: 13–16

Shine HD, Readhead C, Popko B, Hood L, Sidman RL (1992) Morphometric analysis of normal, mutant, and transgenic CNS: Correlation of myelin basic protein expression to myelinogenesis. J Neurochem 58: 342–349

Stoffel W, Boison D, Büssow H (1997) Functional analysis in vivo of the double mutant mouse deficient in both proteolipid protein (PLP) and myelin basic protein (MBP) in the central nervous system. Cell Tiss Res 289: 195–206

Yan Y, Lagenaur C, Narayanan V (1993) Molecular cloning of M6: Identification of a PLP/DM20 gene family. Neuron 11: 423–431

Author's address: Dr. Monika Bradl, Max-Planck-Institut für Neurobiologie, Abteilung Neuroimmunologie, D-82152 Martinsried, Federal Republic of Germany

Molecular mimicry and multiple sclerosis — a possible role for degenerate T cell recognition in the induction of autoimmune responses

B. Gran[1], **B. Hemmer**[1], and **R. Martin**[1,2]

[1] Cellular Immunology Section, Neuroimmunology Branch, National Institutes of Health, Bethesda, MD, U.S.A.
[2] Department of Neurology, University of Maryland at Baltimore Medical School, Baltimore, MD, U.S.A.

Summary. Multiple sclerosis is an inflammatory demyelinating disease of the central nervous system. The etiology is unknown, but several lines of evidence support the hypothesis that the pathogenesis is mediated by autoreactive T lymphocytes. Molecular mimicry has been proposed as a possible mechanism for the development of an autoimmune response to myelin antigens. According to this model, an immune reaction to self antigens could be initiated by T cells that cross-react with infectious agents that "mimic" the autoantigen, i.e. they share immunologic epitopes. It was previously thought that, in order for a cross-reaction of T cells to two different antigens to occur, a substantial amino acid sequence homology between the two antigens was required. More recent studies on the basic mechanisms of T cell antigen recognition have shown that, at least for some T cell clones, antigen recognition is more "degenerate" and sequence homology is not required for crossreactivity to occur. This article reviews the relevance of these recent advances in basic T cell receptor immunology to the occurrence of autoimmunity in the central nervous system.

Introduction

Multiple sclerosis (MS) is an inflammatory demyelinating disease of the central nervous system (CNS). Although the etiology is still unknown, the pathogenesis has long been considered to be mediated by a T cell response against CNS myelin (Martin et al., 1992). Molecular mimicry is one of the proposed mechanisms by which a T-cell response initiated by a foreign pathogen may cause damage to myelin via cross-reactivity with shared immunologic epitopes (Oldstone, 1987). This review summarizes the current evidence for the immunopathogenesis of MS and focuses on recent studies of antigen recognition by T lymphocytes that provide new insight into the possible mechanisms for the initiation and maintenance of autoimmune damage to nervous tissue.

Pathogenetic concepts in multiple sclerosis and experimental autoimmune encephalomyelitis

Evidence for immune-mediated damage of CNS myelin in MS

MS is characterized by relapsing or progressive neurologic deficits that are caused by focal damage to CNS white matter. The pathology of the lesions is characterized by perivascular infiltrates of lymphocytes and macrophages, suggesting that the inflammatory damage is immunologically mediated. Further evidence for an immune pathogenesis is provided by the disease association with the immunogenetic background — in particular certain alleles of the major histocompatibility complex (MHC, HLA in humans), but also other genes — from similarities to the animal model of the disease, experimental autoimmune encephalomyelitis (EAE), and the response to immunosuppressive and immunomodulatory treatments (Martin et al., 1992).

Importance of genetic factors

The prevalence of MS is higher in regions with temperate than in those with tropical climates (Kurtzke, 1985). This difference has been attributed to both environmental factors — such as infectious agents (Kurtzke, 1993) and socioeconomic level — and to genetic factors. Population studies have supported the concept that genetic factors may explain the differences in disease prevalence among distinct ethnic groups living in the same geographic areas. Family and twin studies also support the notion that disease susceptibility is influenced by genetic factors (Ebers et al., 1995).

As will be discussed later, genes of the HLA system are involved in the presentation of antigens to T lymphocytes and therefore play a central role in the immune response. In Caucasian patients, MS is associated with certain alleles of MHC class II genes, such as DR15 Dw2 and DQw6 (Vartdal et al., 1989). Other MHC class II alleles are related to disease susceptibility in selected populations, such as the Japanese (low disease prevalence, association with DR2 and DR6; Hao et al., 1992) and Sardinians (high disease prevalence, association with DR4; Marrosu et al., 1988). A role for the immunogenetic background in the susceptibility for MS has been confirmed by the results of large-scale, multicenter genome screenings (Ebers et al., 1996; Haines et al., 1996; Sawcer et al., 1996).

MS is not a monogenic disease, however; it is likely that several genetic loci contribute to disease pathogenesis in a quantitative manner.

Experimental allergic encephalomyelitis

The first studies of an experimental model of inflammatory demyelination followed the observation that some patients developed an acute encephalomyelitis after vaccination with Pasteur rabies vaccine that was initially prepared in nervous system tissue (Remlinger, 1905). Pivotal studies by Rivers suggested that post-vaccinal encephalomyelitis was not caused by rabies vaccine itself, but by a sensitization against inoculated CNS tissue followed by an immune-mediated attack to myelin (Rivers et al., 1933). This was later proven by inducing an acute or relapsing demyelinating disease, EAE, in susceptible inbred animal strains by injection of myelin proteins (such as myelin basic protein, MBP, or proteolipid protein, PLP) or peptides emulsified in complete Freund's adjuvant (CFA; actively induced EAE). It has subsequently been shown that the disease can be transferred to healthy recipient animals by activated, myelin-specific, CD4+ T cells (passively induced or adoptive transfer EAE), but not by myelin-specific CD8+ T cells or antibodies (Martin et al., 1992). Transgenic mouse models in which the T cell receptor (TCR) of an encephalitogenic T cell clone was expressed as transgene added further evidence to the concept that EAE is a T-cell mediated autoimmune disease (Goverman et al., 1993). Interestingly, encephalitogenic T cell populations can be generated from healthy animals by repeated in vitro stimulation with MBP, documenting that potentially damaging T cells are part of the physiological T cell repertoire (Schlüsener and Wekerle, 1985). The observation that disease susceptibility in different rodent strains is largely under the control of the MHC-class II type of the animal is consistent with the above described association of MS with certain alleles of the HLA class II system in humans (Fritz et al., 1985; Kuokkanen et al., 1996). Although no single EAE model reproduces the complexity of the human immune-mediated demyelination, different animal models may help to understand specific aspects of the human disease, such as the type of clinical course, and the pathology and distribution of white matter lesions (Raine, 1997).

Pathogenetic events in the periphery versus target tissue

The study of EAE has provided tools to define and dissect a series of steps in the development of inflammatory autoimmune damage to CNS tissue. A fundamental characteristic of EAE is that sensitization to myelin antigens used to induce the disease takes place in the periphery, i.e. in lymph nodes draining the immunization site. From regional lymph nodes activated myelin antigen-specific T cells migrate to the systemic circulation, interact with cerebrovascular endothelium and are able to cross the blood-brain barrier (BBB) (Cannella et al., 1990). The expression of adhesion molecules that interact with complementary molecules on the surface of endothelial cells allows activated T cells to adhere to postcapillary venule walls and migrate to the brain parenchyma. Accordingly, specific inhibition of such interactions by monoclonal antibodies has been effective in preventing the induction of EAE (Yednock et al., 1992). After crossing the BBB, myelin specific T cells may have access to the antigen(s) they were primed to recognize in the periphery. Upon recognition of myelin antigens on the surface of CNS antigen presenting cells (i.e. microglia and, to a lesser extent, astrocytes) specific T cell clones may be expanded and produce inflammatory cytokines and chemokines, that cause other cell types — such as macrophages and granulocytes — to migrate to the inflammatory site and contribute to further inflammation and tissue damage.

An obvious difference between MS and EAE is that in the human disease there is no peripheral immunization in the presence of adjuvant substances, like in actively induced EAE. However, it is conceptually possible that a more "subtle" form of peripheral immunization takes place in the human disease. Epidemiological studies suggest that environmental factors may be important in the pathogenesis of MS. In particular, viral infections have been associated with MS exacerbations. The "molecular mimicry" hypothesis of autoimmunity (Fujinami and Oldstone, 1985) postulates that the immune response to a foreign (non-self) invading microorganism, such as a virus, can provoke a reaction to myelin (self) antigens if the non-self and the self antigens share immunologic epitopes. A peripheral event, i.e. the recognition of a foreign antigen by T cells, could be followed by a central event, the cross-recognition of autoantigens in the target organ of the disease, i.e. the CNS (Fig. 1).

For the purpose of understanding how cross-recognition of antigens may take place, a brief summary of basic concepts in T cell antigen recognition will follow.

How T cells recognize antigen

In contrast to antibodies, which are able to react with complex protein- or polysaccharide structures either in particulate form or in solution, T lymphocytes recognize short peptide fragments derived from larger proteins in the context of self-MHC/HLA (Germain, 1994; Zinkernagel and Doherty, 1974). Taking a viral infection as an example, T lymphocytes do not respond to the entire virus, but recognize short peptides generated by intracellular proteolytic degradation of the viral proteins by antigen presenting cells (APC). The peptides are loaded onto newly synthesized MHC molecules and then transported to the surface of APC where the complex of self-MHC and antigenic peptide can be recognized by a specific T cell (MHC-restricted antigen recognition). The antigenic peptides have an optimal length that varies between 8 and 11 amino acids (aa) for MHC class I-restricted recognition by CD8+ T cells, and 12 and 24 aa for class II-restricted presentation to CD4+ T cells (Germain and Margulies, 1993). In contrast to antibodies, that recognize the tridimensional structure of antigenic determinants of soluble proteins, T cells rather recognize the linear structure of antigenic peptides bound to the groove of MHC molecules. In the original formulation of the molecular mimicry hypothesis of autoimmunity, it was postulated that the cross-recognition of self and foreign antigens could be explained by sequence homologies (i.e. identical stretches of aa) between different antigenic proteins. This formulation has subsequently evolved in parallel with the development of new models for T cell antigen recognition.

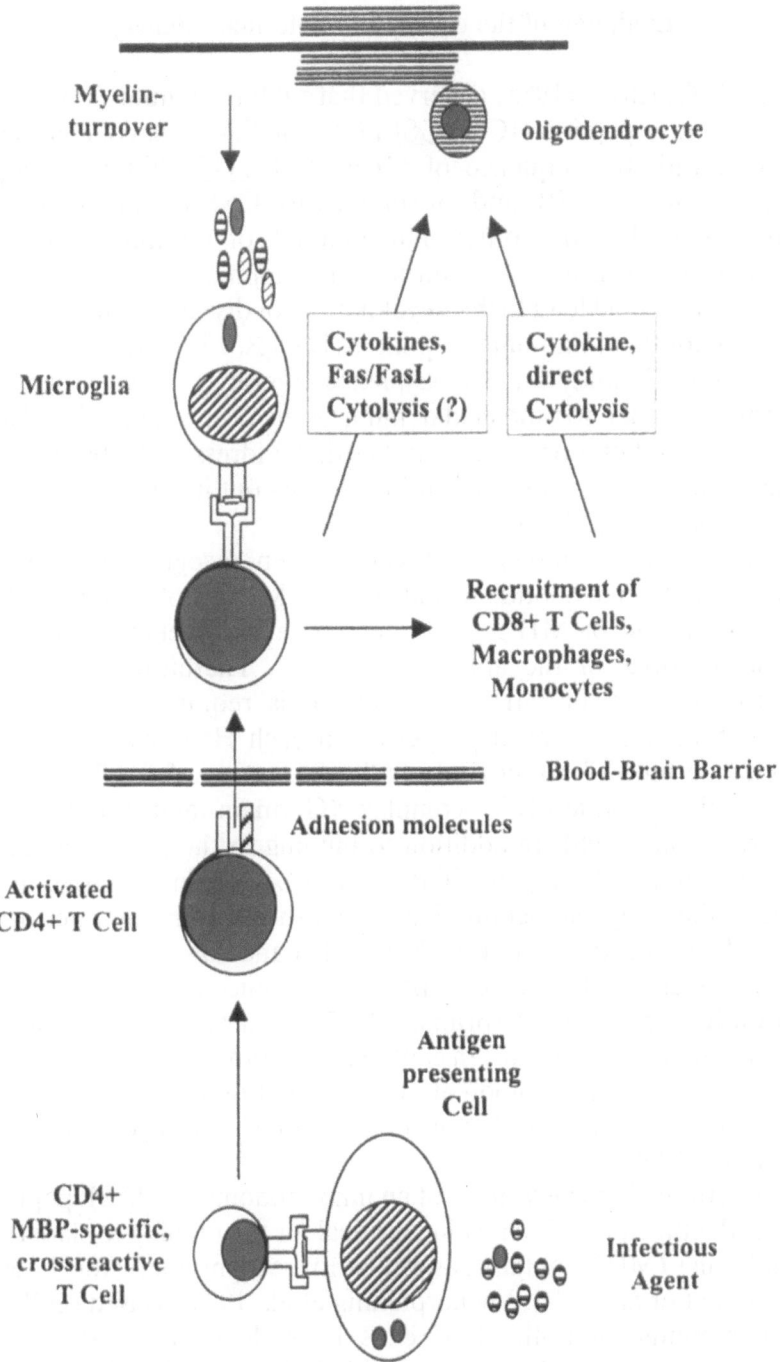

Fig. 1. Proposed mechanism for the occurrence of autoimmunity in the central nervous system. CD4+ T cells are activated in the periphery by antigens derived from infectious agents. These foreign antigens share immunologic epitopes with myelin self antigens. Activated T cells cross the blood-brain barrier and react to antigens derived from myelin, that are presented by resident antigen presenting cells (microglia). The expansion of activated myelin-specific T cell clones leads to autoimmune myelin damage via direct — cytokine production, cytolysis — and indirect — recruitment of other inflammatory cells, antibody help — mechanisms

Evolution of the concept of molecular mimicry

Fujinami and Oldstone (1985) observed that rabbits immunized with a hepatitis B polymerase peptide (ICG<u>YGSLPQ</u>E; one letter aa code is used) that shared 6 aa with the sequence of MBP (TTH<u>YGSLPQ</u>K) developed an antibody response to MBP and, in some cases, CNS lesions reminiscent of EAE. Studies conducted in other animal models of autoimmune diseases — such as adjuvant arthritis — confirmed that mimicry of host antigens by infectious agents could lead to the development of disease by inducing cellular as well as humoral autoimmune responses (Davies, 1997). A crucial aspect of the described EAE study was that sequence homology between two proteins is necessary in order for molecular mimicry to occur. This view has been modified as a result of more recent studies that addressed the basic molecular mechanisms of T cell antigen recognition. A brief outline of important aspects of these mechanisms follows.

The trimolecular complex of T cell antigen recognition is formed by the TCR, the antigenic peptide ligand, and the MHC molecule (Fig. 2a). The floor of the groove of MHC molecules contains "pockets" that favor aa in specific positions of the antigenic peptides. Therefore, in order for a peptide to bind a given MHC molecule, it is required that the peptide bear aa with certain chemical properties in each HLA contact site. These "MHC-binding motifs" are critical for the formation of the T cell receptor ligand, i.e. the peptide-MHC complex (Germain and Margulies, 1993; Rammensee et al., 1995). In addition to binding to the MHC, the antigenic peptide also interacts with the TCR (Fig. 2b). Certain positions in the aa sequence of the antigenic peptide have been shown to be more critical than others for the interaction with the TCR. Allen and coworkers proposed that one key aa residue in the antigenic peptide sequence is strictly necessary for recognition by a given TCR (primary TCR contact) and does not tolerate even conservative substitutions. In contrast, aa in other TCR contact positions (secondary TCR contacts) modulate the recognition and can be substituted with aa that are similar in charge, polarity or size (Evavold et al., 1993; Evavold et al., 1995).

The structural characterization of an immunodominant MBP peptide permitted the identification of aa residues critical for the binding to class II (DR2, DQ1) molecules (MHC contacts) as well as for recognition by the TCR (TCR contacts) (Vogt et al., 1994; Wucherpfennig et al., 1994). Based on the structural requirements for both MHC class II binding and TCR recognition of an immunodominant MBP peptide (MBP (83–97)), Wucherpfennig and Strominger (Wucherpfennig and Strominger, 1995) developed molecular mimicry "motifs" to search a protein database for viral and bacterial peptides that matched those requirements. Of 129 peptides that were identified and synthesized, 8 peptides — 7 of viral and one of bacterial origin — were effective in activating 3 MBP-specific TCC derived from MS patients. A critical observation in this study was that sequence homology was not required for cross-recognition of self and foreign antigens to occur. Indeed, only one of the seven stimulatory peptides could have been identified as a

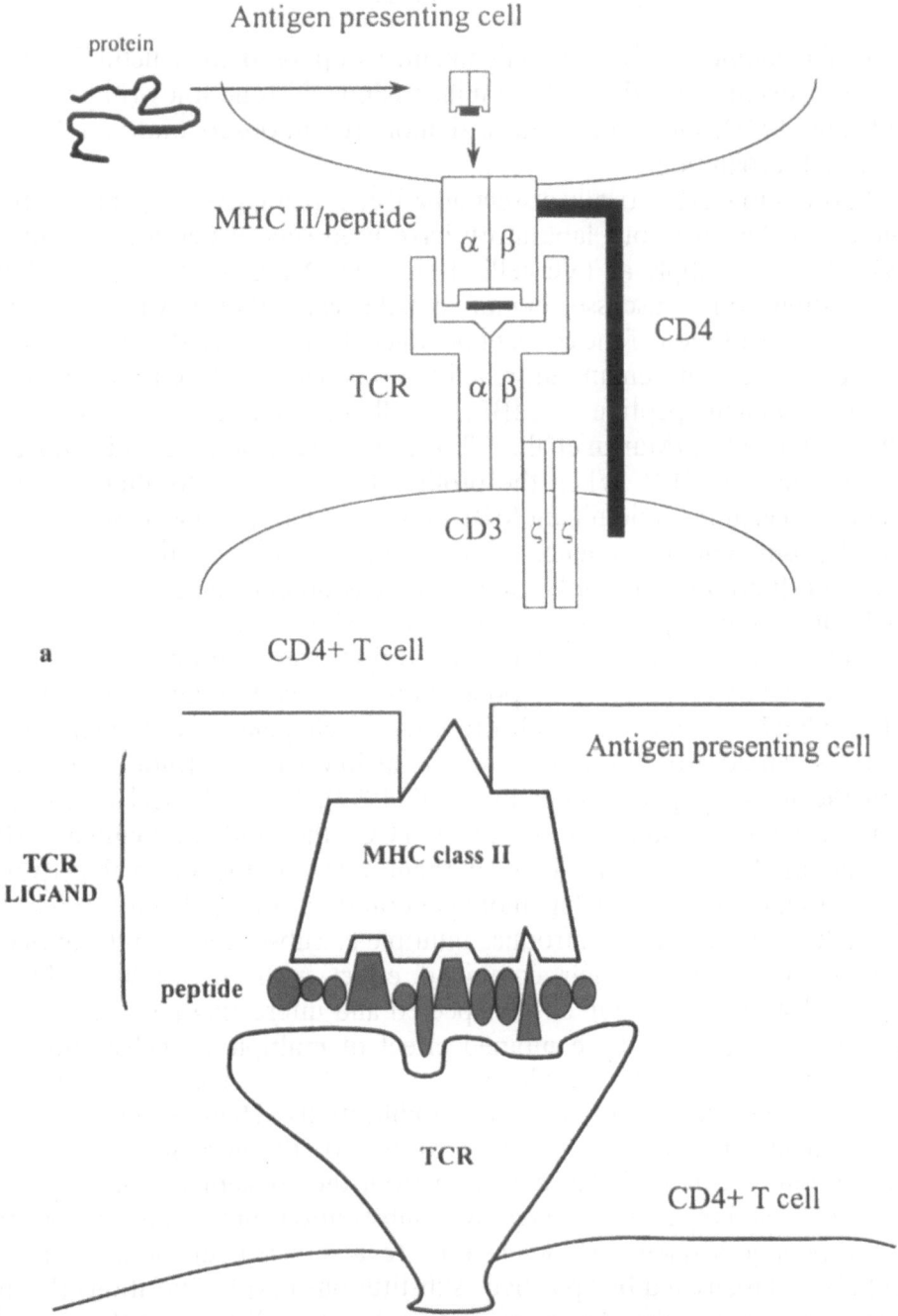

Fig. 2. a CD4+ T cells recognize peptide antigens in the context of self MHC class II molecules. Antigenic proteins are internalized, processed and offered to T cells by antigen presenting cells. CD4 molecules function as coreceptors and are brought into contact with the T cell receptor (*TCR*) complex upon TCR engagement. Signal transduction involves the phosphorylation of ζ chains and other components of the TCR complex (reviewed in Alberola-Ila et al., 1997). **b** The antigenic peptide and the MHC class II molecule form the TCR ligand. For a given antigenic peptide, amino acids in certain positions of the sequence are more important for binding to the MHC (MHC anchors), whereas others preferentially contribute to the interaction with the TCR (TCR contacts). Each amino acid in the sequence, however, can contribute to both interactions. Recent studies suggest that no amino acid in the peptide sequence is strictly necessary for a productive TCR engagement, i.e. the effect of amino acids substitutions in any position of the sequence can be compensated for by single or multiple amino acid substitutions in other positions of the sequence. This observation has led to the identification of agonist ligands with no sequence homology (Hemmer et al., 1998a)

molecular mimic by sequence alignment, as opposed to structural criteria. This is an extension of the earlier Fujinami and Oldstone study (Fujinami and Oldstone, 1985), and is consistent with more recent observations on the rules for T cell recognition.

In order to systematically dissect how T cells recognize antigenic peptides, studies conducted in our laboratory have used sets of peptides substituted at single and multiple aa (Vergelli et al., 1996; Vergelli et al., 1997). Some observations will be discussed in more detail because they may have relevance to a further evolution of the concept of molecular mimicry. Most of the studies were conducted on human autoreactive T cell clones (TCC) specific for an immunodominant peptide of MBP, a well-characterized candidate myelin autoantigen in MS (Martin et al., 1992). TCC were generated and cultured in the presence of MBP. When the proliferative response to altered peptide ligands — i.e. immunodominant MBP peptides bearing single aa substitutions in each position of the sequence — was studied, it was found that, as expected, some substitutions were "tolerated", whereas others caused a reduction or abolition of the response (Vergelli et al., 1996). A rather unexpected and intriguing finding was that certain substitutions were not only tolerated, but actually caused the TCC to respond more effectively than to the "native" MBP peptide. Amino acid substitutions which generated "superagonist" peptides caused a proliferative response at lower concentrations compared with the native peptide (Vergelli et al., 1997). It was logical to conclude that, at least for some autoreactive TCC, the immunodominant MBP peptide used to grow and expand them was not the optimal activating ligand. In the attempt to define more general rules for T cell antigen recognition, a further step was to introduce multiple aa substitutions in the antigenic peptide and compare their combined effect with that induced by the single substitutions. Again, an unexpected and interesting pattern emerged. With few exceptions, the combined effect of multiple aa substitutions on T cell antigen recognition could be predicted by the additive effect — i.e. positive or negative — of single aa modifications (Hemmer et al., 1998a). Interestingly, no single aa seemed to be strictly necessary for antigen recognition — a remarkable deviation from the concept of "primary TCR contact" — and the effect of "negative" substitutions in virtually any position of the peptide sequence (those leading to a reduced functional response) could be compensated by "positive" substitutions in other positions (leading to more potent functional response). As a direct application of these results, it was possible to design peptides that shared no single aa with the native antigenic sequence, and yet were able to stimulate the TCC (Hemmer et al., 1998a).

Even the sequence motifs employed by Wucherpfennig and Strominger may represent a partial view on the concept of T cell receptor degeneracy. In order for cross-recognition of self and non-self antigens to occur, not only is sequence homology not required, but the antigenic peptides of interest need not share a single aa in their sequence. Instead, a certain combination of aa will positively or negatively influence TCR recognition of antigen (Table 1).

Table 1. Evolution of the concept of molecular mimicry. Extent of sequence homology between cross-reactive self peptides and foreign or synthetic peptides

Antigens	Cross-reacting peptides	References
MBP(66–75) HBV-P (589–598)	TTH**YGSLPQ**K ICG**YGSLPQ**E	Fujinami and Oldstone (1985)
MBP(83–97) EBV-P (627–641)	ENP**VVHFFKNIV**TPR TGG**VYHFVKKHVV**HES	Wucherpfennig and Strominger (1995)
MBP(83–97) HSV-T (153–167)	ENPV**VHF**FKNIVTPR FRQL**VHF**VRDFAQLL	Wucherpfennig and Strominger (1995)
MBP(87–99) Predicted peptide	VHFFKNIVTPRTP GGLLAHVISAKKA	Hemmer et al. (1998a)
MBP(89–98) Predicted peptide	FFKNIVTPRT WYALLPSCKG	Hemmer et al. (1998a)

Amino acid sequences of cross-reacting peptides are reported using the one letter code. Sequence homology, i.e. identical amino acids in corresponding positions of cross-reacting peptides, are shown in bold underlined characters. *HBV-P* Hepatitis B virus polymerase; *EBV-P* Epstein-Barr virus DNA polymerase; *HSV-T* Herpes simplex virus terminase

Extensive dissection of TCR antigen recognition by soluble peptide combinatorial libraries

The above described approach examined single and multiple aa substitutions on peptides (or peptide backbones) derived from MBP (83–99). The use of soluble peptide combinatorial libraries (SCLs) as a more unbiased and powerful tool to investigate TCR antigen recognition was instrumental in providing further insight into the degeneracy of TCR-antigen interactions. SCLs are highly complex mixtures of peptides in which the 20 natural aa occur in completely randomized order at each position of the sequence (Houghten et al., 1991; Pinilla et al., 1994). For a given peptide length, they represent the complete set of peptides it is theoretically possible to build. For example, a completely randomized decapeptide library is made of 20^{10} different individual peptides. For the purpose of analyzing T cell antigen recognition, it is also possible to use sets of "sublibraries" that are completely randomized, except for one position in the sequence that bears the same aa, instead of a random mixture of 20. In the example of the 10mer library, it is possible to build 200 "sublibraries" (20 aa × 10 positions in the sequence) each with only one aa fixed in the sequence, whereas all other positions are completely randomized. Such a set of sPCL can be used to test the reactivity of TCC in a "positional scanning" approach (Hemmer et al., 1998b). If a clone preferentially responds to a complex mixture of peptides with one aa fixed in a defined position, that aa is likely to be optimal in that position in a hypothetical antigenic peptide. Using this approach, positional scanning SCL were used to extensively define the spectrum of stimulatory ligands for an autoreactive,

MBP-specific TCC and to search for cross-reactive, high potency ligands (Hemmer et al., 1997). Optimal aa were defined for each position of the antigenic peptide. On the basis of the library predictions, high potency peptides were synthesized, that were found to be effective agonists at concentrations up to 5 orders of magnitude lower that the immunodominant MBP peptide recognized by the TCC. In addition, screening of protein databases with the library information permitted to identify several potential stimulatory peptides derived from both foreign and self antigens. Again, these peptides were synthesized and proved to be effective agonist ligands for the TCC, in some cases at lower concentrations than MBP(86–96). A number of interesting conclusions can be drawn on the basis of these findings; a) TCR recognition appears to be more degenerate that previously thought, at least for some autoreactive TCC; b) the autoantigen that was used to establish and expand the TCC was actually a suboptimal ligand; c) foreign and self peptides can be found that are more potent agonists than MBP, and d) further evidence supports the hypothesis that each aa in the peptide sequence contributes independently to TCR recognition.

Conclusion: Relevance to the occurrence of autoimmunity

Although a detailed discussion of the immunological implications of TCR degeneracy is beyond the scope of this article (the reader is referred to Hemmer et al., 1998b), it should be noted that the concept of degenerate TCR recognition may have relevance not only for the occurrence of autoimmunity, but also for the process of thymic selection — that shapes the mature immune repertoire — and that these two phenomena may be closely related. The repertoire of mature, circulating T cells is formed by the positive selection of those cells that are able to recognize self MHC molecules together with self peptides (von Boehmer, 1994). Negative selection limits the T cell repertoire by eliminating cells with extreme levels of affinity for self MHC-peptide complexes — a) deletion of cells with very high affinity for self MHC-peptide b) lack of expansion and "death by neglect" of cells with insufficient affinity (Nossal, 1994). An interesting hypothesis is that at least part of the positively selected T cell repertoire may be characterized by a high degree of degeneracy in antigen recognition. This part of the repertoire could have a higher intrinsic potential for recognizing both self antigens and cross-reactive foreign antigens, and could be activated by the latter in the course of an infection. The rare occurrence of autoimmunity, however, could be due to the lack of simultaneous facilitating factors — costimulation, opening of the BBB, upregulation of MHC molecules, full as opposed to partial activation of autoreactive T cells — as well as the existence of several levels of protective mechanisms, theoretically including the presence of degenerate, cross-reactive T cells with an inhibitory function (Barnaba and Sinigaglia, 1997).

In conclusion, the data summarized in this article suggest that the conceptual framework for molecular mimicry and the occurrence of autoimmunity should be reformulated. The requirement for cross-activation of autoreactive

TCC by foreign pathogens may follow different rules and may be less stringent that previously thought. It should also be emphasized that crossreactivity is likely to be a normal phenomenon in immune recognition, and may not explain the occurrence of autoimmune diseases by itself. It remains to be proven that an infection is an effective trigger in vivo for self reactive T cells that are part of the normal T cell repertoire.

References

Alberola-Ila J, Takaki S, Kerner JD, Perlmutter RM (1997) Differential signaling by lymphocyte antigen receptors. Annu Rev Immunol 15: 125–154

Barnaba V, Sinigaglia F (1997) Molecular mimicry and T cell-mediated autoimmune disease. J Exp Med 185: 1529–1531

Cannella B, Cross AH, Raine CS (1990) Upregulation and coexpression of adhesion molecules correlate with relapsing autoimmune demyelination in the central nervous system. J Exp Med 172: 1521–1524

Davies JM (1997) Molecular mimicry: can epitope mimicry induce autoimmune disease? Immunol Cell Biol 75: 113–126

Ebers GC, Sadovnick AD, Risch NJ (1995) A genetic basis for familial aggregation in multiple sclerosis. Canadian Collaborative Study Group. Nature 377: 150–151

Ebers GC, Kukay K, Bulman DE, Sadovnick AD, Rice G, Anderson C, Armstrong H, Cousin K, Bell RB, Hader W, Paty DW, Hashimoto S, Oger J, Duquette P, Warren S, Gray T, O'Connor P, Nath A, Auty A, Metz L, Francis G, Paulseth JE, Murray TJ, Pryse-Phillips W, Risch N, et al (1996) A full genome search in multiple sclerosis. Nat Genet 13: 472–476

Evavold BD, Sloan-Lancaster J, Allen PM (1993) Tickling the TCR: selective T-cell functions stimulated by altered peptide ligands. Immunol Today 14: 602–609

Evavold BD, Sloan-Lancaster J, Wilson KJ, Rothbard JB, Allen PM (1995) Specific T cell recognition of minimally homologous peptides: evidence for endogenous ligands. Immunity 2: 655–663

Fritz RB, Skeen MJ, Jen-Chou CH, Garcia M, Egorov IK (1985) Major histocompatibility complex-linked control of the murine immune response to myelin basic protein. J Immunol 134: 2328–2332

Fujinami RS, Oldstone MBA (1985) Amino acid homology between the encephalitogenic site of myelin basic protein and virus: mechanism for autoimmunity. Science 230: 1043–1045

Germain RN (1994) MHC-dependent antigen processing and peptide presentation: providing ligands for T lymphocyte activation. Cell 76: 287–299

Germain RN, Margulies DH (1993) The biochemistry and cell biology of antigen processing and presentation. Annu Rev Immunol 11: 403–450

Goverman J, Woods A, Larson L, Weiner L, Hood L, Zaller DM (1993) Transgenic mice that express a myelin basic protein-specific T cell receptor develop spontaneous autoimmunity. Cell 72: 551–560

Haines JL, Ter-Minassian M, Bazyk A, Gusella JF, Kim DJ, Terwedow H, Pericak-Vance MA, Rimmler JB, Haynes CS, Roses AD, Lee A, Shaner B, Menold M, Seboun E, Fitoussi RP, Gartioux C, Reyes C, Ribierre F, Gyapay G, Weissenbach J, Hauser SL, Goodkin DE, Lincoln R, Usuku K, Oksenberg JR et al (1996) A complete genomic screen for multiple sclerosis underscores a role for the major histocompatability complex. Nat Genet 13: 469–471

Hao Q, Saida T, Kawakami H, Mine H, Maruya E, Inoko H, Saji H (1992) HLAs and genes in Japanese patients with multiple sclerosis: evidence for increased frequencies of HLA-Cw3, HLA-DR2, and HLA-DQB1*0602. Hum Immunol 35: 116–124

Hemmer B, Fleckenstein BT, Vergelli M, Jung G, McFarland H, Martin R, Wiesmueller KH (1997) Identification of high potency microbial and self ligands for a human autoreactive class II-restricted T cell clone. J Exp Med 185: 1651–1659

Hemmer B, Vergelli M, Gran B, Ling N, Conlon P, Pinilla C, Houghten R, McFarland HF, Martin R (1998a) Cutting edge: predictable TCR antigen recognition based on peptide scans leads to the identification of agonist ligands with no sequence homology. J Immunol 160: 3631–3636

Hemmer B, Vergelli M, Pinilla C, Houghten R, Martin R (1998b) Probing degeneracy in T-cell recognition using peptide combinatorial libraries. Immunol Today 19: 163–168

Houghten RA, Pinilla C, Blondelle SE, Appel JR, Dooley CT, Cuervo JH (1991) Generation and use of synthetic peptide combinatorial libraries for basic research and drug discovery. Nature 354: 84–86

Kuokkanen S, Sundvall M, Terwilliger JD, Tienari PJ, Wikstrom J, Holmdahl R, Pettersson U, Peltonen L (1996) A putative vulnerability locus to multiple sclerosis maps to 5p14–p12 in a region syntenic to the murine locus Eae2. Nat Genet 13: 477–480

Kurtzke J (1985) Epidemiology of multiple sclerosis. In: Vinken PJ, Bruyn GB, Klawans HL, et al (eds) Handbook of clinical neurology, vol 3: Demyelinating diseases. Elsevier, Amsterdam, pp 259–287

Kurtzke JF (1993) Epidemiologic evidence for multiple sclerosis as an infection. Clin Microbiol Rev 6: 382–427

Marrosu MG, Muntoni F, Murru MR, Spinicci G, Pischelda MP, Goddi F, Cossu P, Pirastu M (1988) Sardinian multiple sclerosis is associated with HLA-DR4: a serological and molecular analysis. Neurology 38: 1749–1753

Martin R, McFarland HF (1997) Immunology of multiple sclerosis and experimental allergic encephalomyelitis. In: Raine CS, McFarland HF, Tourtellotte WW (eds) Multiple sclerosis: clinical and pathogenetic basis. Chapman Hall, London, pp 221–242

Martin R, McFarland HF, McFarlin DE (1992) Immunological aspects of demyelinating diseases. Annu Rev Immunol 10: 153–187

Nossal GJ (1994) Negative selection of lymphocytes. Cell 76: 229–239

Oldstone MB (1987) Molecular mimicry and autoimmune disease. Cell 50: 819–820

Pinilla C, Appel JR, Houghten RA (1994) Investigation of antigen-antibody interactions using a soluble, non- support-bound synthetic decapeptide library composed of four trillion ($4 \times 10(12)$) sequences. Biochem J 301: 847–853

Raine CS (1997) The lesion in multiple sclerosis and chronic relapsing experimental allergic encephalomyelitis: a structural comparison. In: Raine CS, McFarland HF, Tourtellotte WW (eds) Multiple sclerosis: clinical and pathogenetic basis. Chapman Hall, London, pp 243–286

Rammensee HG, Friede T, Stevanovic S (1995) MHC ligands and peptide motifs: first listing. Immunogenetics 41: 178–228

Remlinger J (1905) Accidents paralytiques au cours du traitement antirabique. Ann Inst Pasteur 19: 625–646

Rivers TM, Sprunt DH, Berry GP (1993) Observations on attempts to produce acute disseminated encephalomyelitis in monkeys. J Exp Med 58: 39–53

Sawcer S, Jones HB, Feakes R, Gray J, Smaldon N, Chataway J, Robertson N, Clayton D, Goodfellow PN, Compston A (1996) A genome screen in multiple sclerosis reveals susceptibility loci on chromosome 6p21 and 17q22. Nat Genet 13: 464–468

Schlüsener H, Wekerle H (1985) Autoaggressive T lymphocyte lines recognize the encephalitogenic region of myelin basic protein; in vitro selection from unprimed rat T lymphocyte populations. J Immunol 135: 3128–3133

Vartdal F, Sollid LM, Vandvik B, Markussen G, Thorsby E (1989) Patients with multiple sclerosis carry DQB1 genes which encode shared polymorphic aminoacid sequences. Hum Immunol 25: 103–110

Vergelli M, Hemmer B, Utz U, Vogt A, Kalbus M, Tranquill L, Conlon P, Ling N, Steinman L, McFarland HF, Martin R (1996) Differential T cell activation by altered peptide ligands derived from myelin basic protein peptide (87–99). Eur J Immunol 26: 2624–2634

Vergelli M, Hemmer B, Kalbus M, Vogt A, Ling N, Conlon P, Coligan JE, McFarland HF, Martin R (1997) Modifications of peptide ligands enhancing T cell responsiveness imply large numbers of stimulatory ligands for autoreactive T cells. J Immunol 158: 3746–3752

Vogt AB, Kropshofer H, Kalbacher H, Kalbus M, Rammensee HG, Coligan JE, Martin R (1994) Ligand motifs of HLA-DRB5*0101 and DRB1*1501 molecules delineated from self-peptides. J Immunol 153: 1665–1673

von Boehmer H (1994) Positive selection of lymphocytes. Cell 76: 219–228

Wucherpfennig KW, Strominger JL (1995) Molecular mimicry in T cell-mediated autoimmunity: viral peptides activate human T cell clones specific for myelin basic protein. Cell 80: 695–705

Wucherpfennig KW, Sette A, Southwood S, Oseroff C, Matsui M, Strominger JL, Hafler DA (1994) Structural requirements for binding of an immunodominant myelin basic protein peptide to DR2 isotypes and for its recognition by human T cell clones. J Exp Med 179: 279–290

Yednock TA, Cannon C, Fritz LC, Sanchez-Madrid F, Steinman L, Karin N (1992) Prevention of experimental autoimmune encephalomyelitis by antibodies against alpha 4 beta 1 integrin. Nature 356: 63–66

Zinkernagel RM, Doherty PC (1974) Restriction of in vitro T cell-mediated cytotoxicity in lymphocytic choriomeningitis within a syngeneic or semiallogeneic system. Nature 248: 701–702

Authors' address: R. Martin, M.D., Cellular Immunology Section, Neuroimmunology Branch, NINDS, National Institutes of Health, Building 10, Room 5B-16, 10 Center Dr MSC 1400, Bethesda, MD 20892, U.S.A.

Dopamine agonists: what is the place of the newer compounds in the treatment of Parkinson's disease?

O. Rascol

Department of Clinical Pharmacology, INSERM U455 and Clinical Investigation Center, University Hospital, Toulouse, France

Summary. Three new dopamine agonists (cabergoline, pramipexole, ropinirole) have been put on to the market within the past months to treat patients with Parkinson's disease. Like any marketed dopamine agonists, the new compounds bind to the D2-like receptors. Pramipexole and ropinirole appear to be quite close drugs. Both are selective non ergot D2 (and preferentially D3) agonists, with an elimination half-life of 5 to 10 hours. Conversely, cabergoline is an ergot derivative, less selective for the D2 receptors, with a much longer elimination half-life (60 hours or more).

In moderately advanced levodopa treated patients with Parkinson's disease and motor fluctuations, cabergoline, pramipexole and ropinirole all do significantly better than placebo in reducing UPDRS motor examination scores, time spent off and daily dose of levodopa. None of the 3 newer agonists proved to do significantly better than bromocriptine in this indication, at the cost of very similar adverse effects.

In de novo levodopa naive patients, pramipexole and ropinirole did significantly better than placebo in short-term (few months) follow-up trials, at the cost again of classical dopaminergic adverse effects. Ropinirole was marginally more effective than bromocriptine, while its use induced the same risk of psychosis than the "old" reference agonist. Early treatment with cabergoline, compared with levodopa, in a long-term (5 year) study reduced the relative risk of developping motor complication by more than 50%. A similar study is presently on-going to compare ropinirole and levodopa. Clinical trials to assess putative neuroprotective effects are also on going with ropinirole and pramipexole. Up to now, the available clinical controlled data suggest that the newer dopamine agonists have very similar clinical effects with only minor superiority, if any, versus bromocriptine.

Introduction

Over the past 40 years, few topics in clinical neuropharmacology have surpassed Parkinson's disease regarding our progresses in understanding the pathophysiological mechanisms of the disease and our pharmacological therapeutic management of the patients. In the late 1960s, the discovery of the dopaminergic nigrostriatal deficit in patients with Parkinson's disease led to the introduction of levodopa therapy. Levodopa still remains, 40 years later, the "gold standard" of any antiparkinsonian treatment (Birkmayer and Hornykiewicz, 1961). However, the initial therapeutic success of levodopa is often blunted within few years of treatment by the development of various motor (fluctuations, abnormal movements) and psychiatric (confusion, hallucinations) side effects (Nutt, 1990; Marsden et al., 1987). Because of such levodopatherapy limitations, the treatment of patients with Parkinson's disease has expanded to incorporate other complementary pharmacological approaches. Among these new therapeutic options, the family of dopamine D2 agonists is one of the most widely used.

Bromocriptine was first introduced as add-on therapy to levodopa in parkinsonian patients experiencing motor fluctuations in the late 1970s (Calne et al., 1974). Seven D2 dopamine agonists are now currently marketed in many countries for the treatment of Parkinson's disease. Four of them are relatively "old" ones: bromocriptine, lisuride, pergolide, piribedil. The three others are "newer" drugs, which have been put on the market within the last twelve months: cabergoline, pramipexole and ropinirole. All these dopamine agonists share the property to bind to the D2-like family of dopamine receptors. However, all of them have specific characteristics, which provide some differences from one drug to the other. Such differences are mainly related to each compound's individual pharmacological profile, in term of receptor selectivity, pharmacokinetics and putative neuroprotective effects (Montastruc et al., 1993; Uitti and Ahlskog, 1996). These differences are often put forward for marketing purposes, to try to establish that these compounds are not "me-too" drugs. However, the results of the controlled comparative clinical trials, when available, are far less demonstrative and do not often convince that any agonist is indeed truely superior to the others.

Which agonist to prefer, and when to use it to treat optimally our patients with Parkinson's disease are thus two distinct questions which both remain unslolved at the present time, because we still lack, in 1998, objective, reliable and convincing demonstrative comparative data. A lot of work, in term of modern Clinical Pharmacology, drug's assessment methods and "Evidence Based Medicine" is still required before anybody can provide unquestionable recommandations in the field.

There are many undisputable pharmacological differences among the marketed D2 agonists. Nobody knows, however, if such properties truly induce relevant differences in term of patients clinical outcome

Looking at the selectivity of the various dopamine agonists for the dopamine receptors, one can notice that those which belong to the ergot derivative family (bromocriptine, lisuride, pergolide, cabergoline) can be seen as so-called "dirty" drugs, because they also have high to moderate affinity for a variety of non-dopaminergic receptors, like the alpha adrenergic and seroton-

Table 1. Affinity of different dopamine agonists for different types of receptors (*D1-like* D1 family of dopamine receptors, *D2-like* D2 family of dopamine receptors, *5HT* serotonergic receptors, *Alpha* alpha adrenergic receptors)

	D1-like	D2-like	5HT	Alpha
Bromocriptine	±	+++	+	+
Lisuride	+	+++	+	+
Pergolide	+	+++	+	+
Cabergolide	+	+++	+	+
Ropinirole	0	+++	0	0
Pramipexole	0	+++	0	0

ergic ones (Piercey et al., 1996; Piercey, 1998; Montastruc et al., 1993; Fariello, 1998) (Table 1). The "older" compounds are then less selective than some "newer" ones, like ropinirole and pramipexole, which are known to exclusively bind to dopamine receptors (Tullosh, 1997; Piercy, 1998). But what does this difference really mean in term of clinical outcomes? A drug acting on noradrenergic or serotonergic receptors might theoretically induce better antiparkinsonian efficacy or have additional useful therapeutic properties (antidepressant, for example). It could also be responsible for more adverse events, like psychosis or dysautonomic symptoms. Clinical studies have not provided yet any controlled data to support the reality of these hypothesis. It is nevertheless likely (although not yet objectively proved), that, in term of adverse events, non ergot derivatives, like pramipexole and ropinirole, will probably induce fewer fibrosis than the ergot compounds, including cabergoline (Bhatt et al., 1991; Frans et al., 1992).

Dopamine agonists also differ among them regarding their relative selectivily and affinity for the D1 versus D2-like receptors (Table 1). Pergolide, for example is a D1 and D2 mixed agonist. Bromocriptine is a D2 agonist and a weak D1 antagonist (partial agonist). Other agonists, like ropinirole and pramipexole, are presented as pure or selective D2 (and preferentially D3) agonists (Piercey, 1998; Tullosh, 1997). Such differences might also have some important clinical correlates. D1 and D2 synergistic activation might be necessary to induce a full antiparkinsonian effect (Robertson, 1992; Luquin et al., 1994). Conversely, D1 receptors have been implicated in the pathophysiological models of several levodopa-induced adverse events, like dyskinesias or psychosis. For example, it has been claimed that the antidyskinetic properties of clozapine might be due to its D1 antagonist effect (Bennett et al., 1994). Levodopa might thus induce dyskinesias via D1 rather than D2 mechanisms. According to this hypothesis, selective D2 agonists might have less propensity than levodopa or mixed agonists to induce dyskinesias. However, it has been shown that a selective D2 agonist, like PHNO, can induce dyskinesia even in the levodopa naive non-primed monkeys (Gomez-Mancilla and Bedard, 1992; Luquin et al., 1994). Conversely, some selective D1 agonists, like A-86929 or ABT 431, seem less likely than D2 agonists to reproduce the levodopa-induced dyskinesias in the primed animals or patients (Grondin et al., 1997; Rascol et al., 1997). Recent data, indeed, support the concept that the pharmacological activations of D1 receptors is not mandatory for production of dyskinesia (Blanchet et al., 1997). In fact, the induction of D3 receptor expression has been recently implicated in the mechanism of behavioral sensitization to levodopa, a model of dyskinesia in the rodent (Bordet et al., 1997). Other speculations have also been proposed regarding the preferential selectivity of some agonists on the D3 receptor. The pathways where the highest concentrations of D3 mRNA are located are the mesolimbic pathways, thought to be involved in motivation (Wise and Rompre, 1989). At least some clinically effective antidepressants (nomifensine) are thought to exert their effects through enhancement of dopamine neurotransmission. This has lead some authors to suggest that pramipexole and ropinirole could have antidepressant effects. This is of significant consideration because depression is a common psychiatric feature of Parkinson's disease (Cummings, 1992). Some psychopharmacological experiments in animals suggest that D3 agonists might indeed have some antidepressive properties (Maj et al., 1997). Pilot open studies have also provided preliminary data in human patients (Szegedi et al., 1997). Once again, however, although such speculations are quite interesting matters of investigation, they remain, up to now, purely theoretical, since no objective undisputable clinical comparative controlled data have ever proved that they lead to relevant differences in the clinical outcome of the patients with Parkinson's disease.

Another matter of interest, when one compares the different D2 agonists, are their pharmacokinetic properties. Dopamine agonists are known to differ among them because of specific pharmacokinetic profiles. Most interest has been paid to their respective plasma-elimination half-life. Unfortunately, we often lack precise data concerning the "older" dopamine agonists half-life, because these drugs have been developped at a time when dosages were usually not very sensitive and registration authorities were not requiring such informations. Consequently, pergolide's half-life remains unknown and has been reported to range from 3 to 27 hours! Bromocriptine is said to have an elimination half-life of 3–8 hours, while that of lisuride would be 2–6 hours. More reliable data are available for the "newer" agonists. Ropinirole's half-life is about 6 hours (Rascol, 1997). That of pramipexole is about 10 hours (Wright et al., 1997) and that of cabergoline is clearly longer, 65 to 110 hours (Fariello, 1998). These differences might have important clinical consequences. For example, the long half-life of cabergoline allows to prescribe it only once a day, which is a quite simple and confortable regimen for an antiparkinsonian drug (Rinne et al., 1998). However, it is also possible that with such a long half-life, troublesome adverse events, like psychosis, might persist for a longer time than with other agonists, after the drug has been washed-out. Theoretically, a dopamine drug with a long half-life is expected to induce a smooth and continuous stimulation of dopamine receptors which is supposed to better mimic the physiological endogenous release of dopamine than the short-lived and pulsatile effects of L-dopa. This last non-physiological pulsatility might be responsible for the occurence of several complications, specially motor fluctuations and dyskinesias (Chase, 1998). At the moment, ropinirole and bromocriptine have proved to induce significantly less dyskinesia than levodopa, while similarly improving motor performance, in the non-primed MPTP-treated monkey (Bedard et al., 1986; Pearce et al., 1998). Cabergoline has also probably a lesser tendency than levodopa to produce dyskinesia in the MPTP monkey model (Grondin et al., 1966). This interesting property might be due to the longer elimination half-life of these agonists compared with that of levodopa. To our best knowledge, this property has not been investigated, or such results have not been published yet, with other agonists. At the moment, there is still very little definite clinical evidence demonstrating that differences in elimination half-life are clearly correlated with the risk of dopaminergic adverse events, like dyskinesias in humans. Prospective comparative clinical trials are now on-going with different D2 agonists to try to answer to this question. The "FIRST" study, comparing the risk of motor complications after 5 years of treatment with a standard and a controlled-release formulation of levodopa failed however to provide supportive clinical results to this hypothesis (Block et al., 1997).

Finally, one must consider, in this "theoretical" part of our discussion, the potential neuroprotective properties of dopamine agonists. Current concepts of the pathogenesis of Parkinson's disease center on the formation of reactive oxygen species and the onset of oxidative stress leading to oxidative damage to substantia nigra pars compacta (Jenner and Olanow, 1996). There are several ways in which it has been speculated that L-dopa might be toxic in vivo (Olanow, 1990): decarboxylation to dopamine and then to toxic metabolits; auto-oxidation to toxic semiquinones and then to reactive oxygen species; reactions with glutathione to form an adduct that depletes reduced glutathione. There are also some evidences that L-dopa has some toxic effects in vitro: it increases lipid peroxidation and hydroxyl radical formation, and it induces apoptosis and mitochondrial respiratory brain damage in cell cultures. However, the toxicity of L-dopa in vivo has been challenged by several authors (Murer et al., 1998) and there are no clinical objective data to support that levodopa is indeed toxic and accelerates disease progression in human patients (Agid, 1998). There are several reasons why a dopamine agonist might be neuroprotective. It could reduce the amount of dopamine released in the synapse by stimulating presynaptic receptors, reducing the neurone's firing rate and thus reducing turn over of dopamine (Sethy et al., 1997). Moreover, if dopamine is indeed neurotoxic, levodopa may then increase the phenomenon, while the use of a dopamine agonists could reduce it by delaying the need for L-dopa and reducing its daily dose, thus reducing a patient's life time dose of L-dopa. These concerns are however relevant only if one accepts the hypothesis that levodopa is indeed toxic, which has not been proved yet. More interestingly, some dopamine agonists, like bromocriptine and pergolide, have free radical scavenging activity (Yoshikawa et al., 1994). Pramipexole has a low oxidation potential, which could also confer free radical scavenging activity on this compound (Hall et al., 1996). In vivo, pergolide is the only dopamine agonist, to our best knowledge, that has been shown to improve by oral route age-related loss of dopamine neurons in the rat (Felton et al., 1992). All these indirect experimental arguments keep the hypothesis of a neuroprotective role for dopamine agonists, going on. There is the hope that these drugs might then slow down the disease's progression. However, up to now, none of the available dopamine agonist has ever proved that such exiting, but purely speculative, effects have true clinical correlates. Clinical trials using clinical and neuroimaging out-comes are presently on-going with pergolide, pramipexole and ropinirole to test this hypothesis. We must wait for their results before accepting any conclusion. Very preliminary data have however been recently presented, showing that this might indeed be the case for ropinirole, but larger controlled prospective clinical trials are required before accepting such data (Rakshi et al., 1998).

Evidence Based Medicine and the "newer" D2 agonists in the treatment of Parkinson's disease

Once such theoretical arguments have been summarized, we are now left with the most important part of our discussion: what do the available results of the comparative clinical trials, providing "Evidence-Based Medicine", tell us about the usefulness of dopamine agonists and how do they compare? There are very few data concerning the "older" agonists, because these drugs have been developed in times when the clinical pharmacological methods in the field were not "mature". There are few, if no, available satisfying double-blind comparative controlled studies with these drugs. We will then rather focus on the "newer" agonists: cabergoline, pramipexole and ropinirole. These drugs have been studied with more objective and reliable methods. Several trials, performed in two different situations (advanced Parkinson's disease and early Parkinson's disease), using a double-blind, randomized, controlled, prospective design, have now been published with these 3 compounds.

Advanced Parkinson's disease

In this situation, the dopamine agonist is adjuncted to levodopa when this last drug has already induced motor complications, generally motor fluctuations. Several trials using a quite similar design have recently shown that cabergoline, pramipexole and ropinirole are all doing better than placebo in patients with Parkinson's disease suffering from the wearing-off phenomenon. In a 32 week double-blind placebo-controlled parallel-group study conducted in 360 patients with Parkinson's disease and motor fluctuations, Lieberman and colleagues (1993) showed that pramipexole, up to 4.5 mg/d, significantly improved the mean score of UPDRS part II (22% reduction with pramipexole vs. 4% with placebo). In the same study, UPDRS part III was also significantly more improved with pramipexole than placebo (25% vs. 12% reduction). The time spent off also decreased by 31% (vs. 7% with placebo) and the daily dose of levodopa was decreased by 27% (vs. 5% with placebo). In a 3 month double-blind placebo-controlled parallel-group study conducted in 46 patients with motor fluctuations, ropinirole (up to 4 mg bid) reduced the mean percentage of time spent off by 44% compared to 24% with placebo (Rascol et al., 1996). Ropinirole has also been compared to placebo in a 6 month randomized double-blind parallel-group study conducted in 149 patients with Parkinson's disease not optimally controlled on levodopa. There was a statistically significant treatment difference in favor of ropinirole, with more patients (28%) exhibiting a 20% reduction in levodopa dose and a 20% reduction in awake time spent off, in the active group compared with 11% in the placebo group (Kreider et al., 1996). In 188 patients with Parkinson's disease and motor complications, a 24 week double-blind placebo-controlled parallel-group study showed that cabergoline (up to 5 mg once a day) significantly improved UPDRS part II (19% reduction, vs. 4% with placebo) and UPDRS part III (16% vs. 6%). The percentage of time spent on during the day was also improved by cabergoline while the daily dose of Levodopa was reported to be significantly reduced (Hutton et al., 1996). Taken together, these data show that the 3 newer agonists can improve the UPDRS motor scores by 20–30%. The 3 of them can reduce the day time spent off by a couple of hours and can allow to reduce the levodopa daily dose by about 20%. This is observed at the cost of adverse events which are quite similar to what has already been previously reported with the "older" agonists, worsening of dyskinesias, orthostatic hypotension, nausea and hallucinations being among the most frequent ones.

Pramipexole, ropinirole and cabergoline have also been compared to bromocriptine in 3 separate trials. Guttman and colleagues (1997) studied in a double-blind parallel group trial 247 patients suffering from the wearing-off phenomenon treated either by pramipexole (up to 4.5 mg/d), bromocriptine (up to 30 mg/d) and placebo. Both agonists did better than placebo in term of motor efficacy outcomes (UPDRS parts II and III, percentage of time spent off) but there was no significant difference between the effects of both agonists, although pramipexole antiparkinsonian effects seemed somewhat more pronounced than those of bromocriptine. Adverse events were very similar with both agonists. In a 9 month double-blind paralled-group study conducted in 44 patients with Parkinson's disease and motor fluctuations, Inselberg and colleagues (1996) could not find any difference in efficacy or safety profiles of cabergoline (up to 6 mg/d) and bromocriptine (up to 40 mg daily). Finally in a 6 month double-blind parallel-group trial, ropinirole has also been compared with bromocriptine in 555 levodopa-treated patients. Both agonists had a very similar symptomatic effect in term of efficacy as well as safety (Ropinirole 043 Study Group, 1996). Although the results have sometimes been presented as in favor of the "newer" agents, compared to the "older" reference drug, none of these 3 trials provides any statistically significant difference which could consistently support the superiority of ropinirole, pramipexole or cabergoline over bromocriptine in this indication. Unfortunately, there is, to our best knowledge, no published trial comparing the "newer" agonists among them (cabergoline vs. pramipexole vs. ropinirole). Similarly, up to now, no data is available comparing any "newer" agonist to other "older" agonists, like lisuride or pergolide, which are however often more frequently used than bromocriptine in many countries. Finally, none of the "newer" agonist has been compared to MAO-B inhibitors or to COMT inhibitors in advanced Parkinson disease, while these two classes of drugs are also marketed in the same indication.

Early Parkinson's disease

There are several trials which have been conducted with the "newer" agonists in early untreated patients with Parkinson's disease. Such trials have been designed because pilot open clinical trials with some "older" agonists, like bromocriptine, have suggested that using a dopamine agonist early in the course of the disease allows to delay the need for levodopa, allows to reduce the doses of levodopa which are needed to control the symptoms, and more importantly reduces the risk of long-term motor complications (Rascol et al., 1979; Montastruc et al., 1994). The initial pilot studies assessing this question with the "older" agonists all suffered from evident methodological short-comings (Factor and Weiner, 1993). However, recent animal experiments have shown that, like in humans, some agonists, like ropinirole and bromocriptine, elicit comparable dyskinesia than levodopa once levodopa priming has occured, while they have a marked lesser tendency than levodopa to produce dyskinesia while similarly improving motor performance in drug-naive MPTP-treated monkeys. These results theoretically predict a similar response, at least to ropinirole and bromocriptine, in levodopa-naive patients with Parkinson's disease and emphasize the theoretical importance of avoiding initial dyskinesia induction through early use of dopamine agonist drugs (Pearce et al., 1998).

The development of the "newer" dopamine agonists offered a good opportunity to design and conduct more convincing and appropriate trials than the previous available ones, to assess the true place of dopamine agonists in the treatment of the early stages of the disease. This was particularly interesting because it is still commonly beleived that dopamine agonists can only be maintained as monotherapy in a very small proportion of patients after 2 to 3 years of treatment (less than 5% according to several authors) and that they often induce unacceptable severe psychiatric side effects which markedly reduce their benefit/risk ratio in clinical practice. Pramipexole and ropinirole have both been compared to placebo in several hundreds of "de novo" patients with Parkinson's disease enrolled in a number of prospective controlled double-blind trials (Wheadon et al., 1996; Adler et al., 1997; Brooks et al., 1998; Hubble et al., 1995; Kieburtz et al., 1997; Shannon et al., 1997). All published studies showed that each agonist did better than placebo up to 6 months of treatment, with a reduction in UPDRS baseline motor score ranging from 20 to 40%. In all these studies, the most frequent adverse events were quite similar for both drugs and identical to what is expected with any active dopaminergic compound (digestive, cardiovascular and neuropsychiatric effects). To our best knowledge, there is non published study comparing cabergoline to placebo in de novo patients with Parkinson's disease.

Cabergoline and ropinirole have also been compared to levodopa in long-term (5 year) prospective parallel double-blind randomized studies. The objective of these studies was to assess if the early use of such agonists actually delays the incidence of long-term motor complications. The planned interim analysis of both studies (at 6 and 12 months respectively) have been published (Rascol et al., 1998; Rinne et al., 1997). In both cases, levodopa induced a marginal (10%) but statistically significantly greater improvement in UPDRS score than the agonist. There was no difference in term of adverse events profiles. The long-term results (3 to 5 years) of the cabergoline versus levodopa study are now available (Rinne et al., 1998). Four hundred and forty two patients with early idiopathic Parkinson's disease were randomised to receive either cabergoline (up to 4 mg once daily) or levodopa (up to 600 mg/ d). Open labelled levodopa was added in both treatments if needed during the study. The results show that the development of motor complications was significantly less frequent in patients treated with cabergoline than in levodopa recipients (22% vs. 34%) and that the relative risk of developing motor complication during treatment with cabergoline was more than 50% lower than with levodopa. This trials also show that 35% of the patients treated with cabergoline did not require additional levodopatherapy all along the follow-up, remaining on agonist monotherapy at the time of analysis. Overall, the mean cumulative exposure to levodopa in the cabergoline group was 50% less than in the levodopa group (303 g versus 637 g). The ropinirole versus levodopa study enrolled 282 "de novo" patients with Parkinson's disease. It is now one year before the last visit of the last patient recruited to the trial and we must then still wait several months before any conclusion to be drawn. To our best knowledge, there are no such long-term (5 year) study comparing the outcome of patients treated from the early stages with pramipexole versus levodopa. However, a prospective two-year study is now on-going with this compound, but might be too short to detect truely clinically relevant differences.

Ropinirole has also been compared to bromocriptine in a double-blind parallel-group study conducted in 335 patients with early idiopathic Parkinson's disease. After 6 months, ropinirole did slightly (10%), but significantly better than bromocriptine on UPDRS scores (Korczyn et al., 1998). After 3 years of treatment, and despite the fact that ropinirole is a selective agonist at the D2 (D3) receptors, while bromocriptine also binds to other receptors like the serotonergic ones — which have been claimed to participate in the pathophysiology of various psychiatric syndromes — both drugs induced the same amount of psychiatric adverse events (9% of each group experienced hallucinations). The number of patients considered to have dyskinesia was very low in both treatment groups (5% or less in each group) (Larsen et al., 1998). Interestingly, after 3 years of follow-up, more than 30% of the patients were still treated by an agonist as monotherapy (either ropinirole or bromocriptine), with satisfying clinical improvement and no troublesome adverse event, confirming the assumption that a much larger proportion of early patients with Parkinson's disease than what is still too often claimed by some opinion leaders can be adequately manadged during several years with an agonist alone.

To our best knowledge, none of the "newer" dopamine agonist has ever been compared to selegiline, L-dopa slow release formulations or COMT inhibitors in "de novo" patients.

Conclusion

In summary, pramipexole and ropinirole are 2 drugs which look very much the same. They are both selective D2 (D3) agonists with an intermediate elimination half-life of few hours, definitely longer than that of levodopa. Cabergoline is an ergot derivative, with its consequent effects on non-dopamine receptors, and has the longest elimination half-life (more than 60 years). These 3 "new" dopamine agonists, can reduce L-dopa daily dose, can reduce the percentage of time spent off and can improve UPDRS scores in moderately advanced fluctuating patients with Parkinson's disease. This is achieved, however, with no clear superiority over bromocriptine, and there is no clinical evidence that these new agents have significant distinctive efficacy or adverse events profiles. In early PD, the "newer" agonists do significantly better than placebo in controlling parkinsonian symptoms. These agents can delay the need for levodopa by at least several months in most patients, and in one third of the patients for more than 3 years. These drugs have a comparable profile of peripheral and psychiatric adverse events than L-dopa and "older" agonists on short-term follow-up. The first available results of a 5-year L-dopa-controlled study show that the early use of cabergoline in "de novo" patients with Parkinson's disease significantly reduces the risk of occurence of motor complications (fluctuations, and specially, dyskinesias). This is the first time that the pilot results of randomized prospective open studies conducted with bromocriptine (Montastruc et al., 1994) are reproduced with a more modern and reliable methodology. In the near future, the result of another long-term 5-year L-dopa-controlled study with ropinirole should allow to definitely accept or refute the assumption that the early use of dopamine agonists significantly reduce or delay the risk of motor fluctuations and/or dyskinesias in patients with Parkinson's disease.

References

Adler CH, Sethi KD, Hauser RA, Davis TL, Hammerstad JP, Bertoni J, Taylor RL, Sanchez-Ramos J, O'Brien CF, for the Ropinirole Study Group (1997) Ropinirole for the treatment of early Parkinson's disease. Neurology 49: 393–399

Agid Y (1998) Levodopa: is toxicity a myth. Neurology 50: 858–863

Bedard PJ, Di Paolo T, Falardeau P, Boucher R (1986) Chronic treatment with L-dopa, but not bromocriptine induces dyskinesia in MPTP-parkinsonian monkeys. Correlation with [3H] spiperone binding. Brain Res 379: 294–299

Bennett JP, Landow ER, Dietrich S, Schuh LA (1994) Suppression of dyskinesias in advanced Parkinson's disease: moderate daily clozapine doses provide long-term dyskinesia reduction. Mov Disord 9: 404–414

Bhatt MH, Keenan SP, Fleetham JA, Calne DB (1991) Pleuropulmonary disease associated with dopamine agonist therapy. Ann Neurol 30: 613–616

Birkmayer W, Hornykiewicz O (1961) Der L-3, 4 dioxyphenylalamin (= dopa)-Effekt bei der Parkinson-Akinese. Wien Klin Wochenschr 74: 787–788

Blanchet PJ, Konitsiotis S, Chase TN (1997) Motor response to a dopamine D3 receptor preferring agonist compared to apomorphine in levodopa-primed 1-methyl-4-phenyl-1,2,3,6-tetrahydropyridine monkeys. J Pharmacol Exp Ther 283: 794–799

Block G, Liss C, Reines S, Irr J, Nibbelink D and the CR First Study Group (1997) Comparison of immediate-release and controlled release carbidopa/levodopa in Parkinson's disease. Eur Neurol 37: 23–27

Bordet R, Ridray S, Carboni S, Diaz J, Sokoloff P, Schwartz JC (1997) Induction of dopamine D3 receptor expression as a mechanism of behavioral sensitization to levodopa. Proc Natl Acad Sci USA 94: 3363–3367

Brooks DJ, Abbott RJ, Lees AJ, Martignoni E, Philcox DV, Rascol O, Roos RAC, Sagar HJ (1998) A placebo-controlled evaluation of ropinirole, a novel D2 agonist, as sole dopaminergic therapy in Parkinson's disease. Clin Neuropharmacol 21: 101–107

Calne DB, Teychenne PF, Leigh PN, Baniji AN, Greenacre SK (1974) Treatment of Parkinson's disease with bromocriptine. Lancet 2: 1355–1356

Chase TN (1998) The significance of continuous dopaminergic stimulation in the treatment of Parkinson's disease. Drug 55: S1–S9

Cummings JL (1992) Depression and Parkinson's disease: a review. Am J Psychiatry 149: 443–454

Factor SA, Weiner WJ (1993) Early combination therapy with bromocriptine and levodopa in Parkinson's disease. Mov Disord 3: 257–262

Fariello RG (1998) Pharmacodynamic and pharmacokinetic features of Cabergoline. Rationale for use in Parkinson's disease. Drugs 55: S10–S16

Felton DL, Felton SY, Fuller RW, Romano TD, Smalstig EB, Wong DT, Clemess JA (1992) Chronic dietary pergolide preserves nigrostriatal neuronal integrity in aged Fischer-344 rats. Neurobiol Aging 13: 339–351

Frans E, Dom R, Demedts M (1992) Pleuropulmonary changes during treatment of Parkinson's disease with a long acting ergot derivative, cabergoline. Eur Respir J 5: 263–265

Gomez-Mancilla B, Bedard PJ (1992) Effect of chronic treatment with (+)-PHNO, a D2 agonist in MPTP-treated monkeys. Exp Neurol 117: 185–188

Grondin R, Goulet M, Di Paolo T, Bedard PJ (1996) Cabergoline, a long-acting dopamine D2-like receptor agonist, produces a sustained antiparkinsonian effect with transient dyskinesias in parkinsonian drug-naive primates. Brain Res 735: 298–306

Grondin R, Bedard PJ, Britton DR, Shiosaki K (1997) Potential therapeutic use of the selective dopamine D1 receptor agonist, A-86929: an acute study in parkinsonian levodopa-primed monkeys. Neurology 49: 421–426

Guttman M and the International Pramipexole-Bromocriptine Study Group (1997) Double-blind comparison of pramipexole and bromocriptine treatment with placebo in advanced Parkinson's disease. Neurology 49: 1060–1065

Hall ED, Andrus PK, Oostveen JA, Althans JS, Von Voigtlander PF (1996) Neuroprotective effects of the dopamine D2/D3 agonist pramipexole against post ischemic or metamphetamine induced degeneration of nigrostriatal neurons. Brain Res 742: 80–88

Hubble JP, Koller WC, Certler NR, Sramek JJ, Friedman J, Goetz C, Ranhosky A, Korts D, Elvin A (1995) Pramipexole in patients with early Parkinson's disease. Clin Neuropharmacol 18: 338–347

Hutton JT, Koller WC, Ahlskog JE, Pahwa R, Hurting HI, Stern MB, Hiner BC, Lieberman A, Pfeiffer RF, Rodnitzky RL, Waters CH, Muenter MD, Adler CH,

Morris JL (1996) Multicenter, placebo-controlled trial of cabergoline taken once daily in the treatment of Parkinson's disease. Neurology 46: 1062–1065

Jenner P, Olanow CW (1996) Oxidative stress and the pathogenesis of Parkinson's disease. Neurology 47 [Suppl 3]: S161–S170

Kieburtz K, for the Parkinson Study Group (1997) Safety and efficacy of pramipexole in early Parkinson disease. A randomized dose-ranging study. JAMA 2: 125–130

Korczyn A, Brooks D, Brunt E, Poewe W, Rascol O, Stocchi F, on the behalf of the 053 study group (1998) Ropinirole versus bromocriptine in the treatment of Parkinson's disease: a 6 month interim report of a 3 year study. Mov Disord 13: 46–51

Kreider M, Knox S, Gardiner D, Wheadon D (1996) A multicenter double-blind study of ropinirole as an adjunct to L-dopa in Parkinson's disease. Neurology 46: 475

Larsen JP, Brunt E, Korczyn AD, Nagy Z, Poewe W, Ruggieri S and the 053 Study Group (1998) Ropinirole is effective in long-term treatment of patients with early Parkinson's disease. Neurology 50: 277–278

Lieberman A, Imke S, Muenter M, Wheeler K, Ahlskog JE, Matsumoto JY, Maraganore DM, Wright KF, Schoenfelder J (1993) Multicenter study of cabergoline, a long-acting dopamine receptor agonist, in Parkinson's disease patients with fluctuating responses to levodopa/carbidopa. Neurology 43: 1981–1984

Luquin MR, Laguna J, Obeso JA (1992) Selective D2 receptor stimulation induces dyskinesia in parkinsonian monkeys. Ann Neurol 31: 551–554

Luquin MR, Guillen J, Martinez-Vila E, Laguna J, Martinez-Lage JM (1994) Functional interaction between dopamine D1 and D2 receptors in "MPTP" monkeys. Eur J Pharmacol 253: 215–224

Maj J, Rogoz Z, Skuza G, Kolodziejczyk K (1997) Antidepressant effects of pramipexole, a novel dopamine receptor agonist. J Neural Transm 104: 525–533

Marsden CD, Parkes JD, Quinn N (1987) Fluctuations of disability in Parkinson's disease: pathophysiological aspects. In: Marsden CD, Fahn S (eds) Movement disorders. Butterworth, London, pp 96–122

Molho ES, Factor SA, Weiner WJ, Sanchez-Ramos JR, Singer C, Shulman L, Brown D, Sheldon C (1995) The use of pramipexole, a novel dopamine (DA) agonist, in advanced Parkinson's disease. J Neural Transm 45: 225–230

Montastruc JL, Rascol O, Senard JM (1993) Current status of dopamine agonists in Parkinson's disease management. Drugs 46: 384–393

Montastruc JL, Rascol O, Senard JM, Rascol A (1994) A randomised controlled study comparing bromocriptine to which levodopa was later added, with levodopa alone in previously untreated patients with Parkinson's disease: a five year follow up. J Neur Neurosurg Psychiatry 57: 1034–1038

Murer MG, Dziewczapolski G, Menalled LB, Garcia MC, Agid Y, Gershanik O, Raisman-Vozari R (1998) Chronic levodopa is not toxic for remaining dopamine neurons, but instead promotes their recovery, in rats with moderate nigrostriatal lesions. Ann Neurol 43: 561–575

Nutt JG (1990) Levodopa-induced dyskinesia. Neurology 40: 340–345

Olanow CW (1990) Oxidation reactions in Parkinson's disease. Neurology 40: S32–S39

Pearce RKB, Banerji T, Jenner P, Marsden CD (1998) De novo administration of ropinirole and bromocriptine induces less dyskinesia than L-dopa in the MPTP-treated marmoset. Mov Disord 13: 234–241

Piercey MF (1998) Pharmacology of pramipexole, a dopamine D3-preferring agonist useful in treating Parkinson's disease. Clin Neuropharmacol 21: 141–151

Piercey MF, Hoffmann WE, Smith HW, Hyslop DK (1996) Inhibition of dopamine neuron firing by pramipexole, a D3-preferring dopamine agonist: comparaison to other dopamine agonists. Eur J Pharmacol 312: 35–44

Rakshi JS, Bailey DL, Uema T, Morrish PK, ITO K, Brooks DJ (1998) Is ropinirole, a selective D2 receptor agonist, neuroprotective in early Parkinson's disease? An [18F] dopa PET study. Neurology 50: 330

Rascol O (1997) Ropinirole: clinical profile. In: Olanow CW, Obeso JA (eds) Beyond the decade of the brain: Dopamine agonists in early Parkinson's disease. Wells Medical Ltd., Royal Tunbridge Wells, pp 163–176

Rascol A, Guiraud B, Montastruc JL, David J, Clanet M (1979) Long-term treatment of Parkinson's disease with bromocriptine. J Neurol Neurosurg Psychiatry 42: 143–150

Rascol O, Lees A, Senard JM, Pirtosek Z, Montastruc JL, Fuell D (1996) Ropinirole in the treatment of levo-dopa induced motor fluctuations in patients with Parkinson's disease. Clin Neuropharmacol 19: 234–245

Rascol O, Blin O, Descombes S, Soubrouillard C, Fabre N, Viallet F, Thalamas C, Azulay JP, Lafnitzegger K, Fredrick E, Wright S (1997) ABT 431, a selective D1 agonist has efficacy in patients with Parkinson's disease. Neurology 48: 269–270

Rascol O, Brooks D, Brunt E, Korczyn A, Poewe W, Stocchi F, on the behalf of the 056 study group (1998) Ropinirole in the treatment of early Parkinson's disease: a 6 month interim report of a 5 year L-dopa controlled study. Mov Disord 13: 39–45

Rinne UK, Bracco F, Chouza C, Dupont E, Gershanik O, Marti Masso JF, Montastruc JL, Marsden CD, Dubini A, Grimaldi NOR, the PKD009 Collaborative Study Group (1997) Cabergoline in the treatment of early Parkinson's disease: results of the first year of treatment in a double-blind comparison of cabergoline and levodopa. Neurology 48: 363–368

Rinne UK, Bracco F, Chouza C, Dupont E, Gershanik O, Marti Mosso JF, Montastruc JL, Marsden CD, and the Parkinson's Disease S009 Study Group (1998) Early treatment of Parkinson's disease with cabergoline delays the onset of motor complications. Results of a double-blind levodopa controlled trial. Drugs 55: S23–S30

Robertson HA (1992) Dopamine receptor interactions: some implications for the treatment of Parkinson's disease. Trends Neurosci 15: 201–206

Ropinirole 043 Study Group (1996) A double-blind comparative study of ropinirole vs. bromocriptine in the treatment of Parkinsonian patients not optimally controlled on L-dopa. Mov Disord 11: S188

Sethy VH, Wu H, Oostveen JA, Hall ED (1997) Neuroprotective effects of the dopamine agonists pramipexole and bromocriptine in 3-acetylpyridine-treated rats. Brain Res 754: 181–186

Shannon KM, Bennett JP, Friedman JH, for the Pramipexole Study Group (1997) Efficacy of pramipexole, a novel dopamine agonist, as monotherapy in mild to moderate Parkinson's disease. Neurology 49: 724–728

Szegedi A, Hillert A, Wetzel H, Klieser E, Gaebel W, Benkert O (1997) Pramipexole, a dopamine agonist, in major depression: antidepressant effects and tolerability in an open-label study with multiple doses. Clin Neuropharmacol 20: 536–545

Tulloch IF (1997) Pharmacologic profile of ropinirole: a nonergoline dopamine agonist. Neurology 49: S58–S62

Uitti RJ, Ahlskog JE (1996) Comparative review of dopamine receptor agonist in Parkinson's disease. CNS Drugs 5: 369–388

Wheadon D, Wilson-Lynch K, Gardiner D, Kreider M (1996) The efficacy and safety of ropinirole in early Parkinsonian patients not receiving dopaminergic therapy: a multicenter double-blind study. Neurology 46: 159

Wise RA, Rompre PP (1989) Brain dopamine and reward. Annu Rev Psychol 40: 191–225

Wright CE, Sisson TL, Ichhpurani AK, Peters GR (1997) Steady-state pharmacokinetic properties of pramipexole in healthy volunteers. J Clin Pharmacol 37: 520–525

Yoshikawa T, Minamiyama Y, Naito Y, Kondo M (1994) Antioxidant properties of bromocriptine, a dopamine agonist. J Neurochem 62: 1034–1038

Author's address: Professor O. Rascol, Laboratoire de Pharmacologie Médicale et Clinique, Faculté de Médecine, 37 Allées Jules Guesde, F-31073 Toulouse Cedex, France

Apomorphine has a potent antiproliferative effect on Chinese hamster ovary cells

M. Scarselli[1], **P. Barbier**[1], **F. Salvadori**[1], **M. Armogida**[1], **P. Collecchi**[2], **C. Pardini**[1], **F. Vaglini**[1], **R. Maggio**[1], and **G. U. Corsini**[1]

[1] Department of Neuroscience, University of Pisa, Italy
[2] Department of Oncology, Division of Pathology, University of Pisa, Italy

Summary. Apomorphine is a potent non selective agonist at the D_1 and D_2 dopamine receptors acting both pre- and post-synaptically. In this report we describe a novel function of apomorphine, independent from its dopaminergic activity. Apomorphine inhibits Chinese hamster ovary (CHO)-K1 cell proliferation in a dose-dependent manner. The EC_{50} of apomorphine-induced inhibition of CHO-K1 cell proliferation determined by cell counting was $3.24 \pm 0.07\,\mu M$. Remarkably, the dose-response curve obtained by measuring the incorporation of [^{3}H]thymidine was practically identical to the previous one giving an EC_{50} of $3.52 \pm 0.04\,\mu M$. The dopaminergic antagonists SCH23390 and spiperone at a concentration of $10\,\mu M$ (well beyond their K_d values for the dopamine D_1- and D_2-like receptors respectively) were not able to antagonize the effect of apomorphine on CHO-K1 cell proliferation. Apomorphine exerts its effect early during incubation; CHO-K1 cells exposed to apomorphine for a period as short as 1h and then allowed to grow for three days were significantly reduced in number with respect to untreated control cells. After four hours of exposition to apomorphine ($10\,\mu M$) the antiproliferative effect was similar to that seen when this compound was present in the bath for all three days. Concentrations of apomorphine higher than $10\,\mu M$ induced cell death, and the colony was completely destroyed at $50\,\mu M$. Cytometric analyses showed a significant accumulation of CHO-K1 cells in the G2/M phase.

Introduction

Apomorphine (10,11-dihydroxyaporphine), an aporphine alkaloid, was first synthesized by the rearrangement of morphine in the presence of concentrated hydrochloric acid. Soon after its discovery, it was used by veterinary scientists as an effective therapy for behavioural vices in domesticated animals. It was then employed at different times in clinical medicine as an emetic, an expectorant, a sedative, an aphrodisiac, an anti-psychotic, an anticonvulsant and in the management of narcotic and alcoholic dependence

(Lal, 1988). It was not until the discovery of its mechanism of action that a rational use of the drug began. (Ernst, 1967; Anden et al., 1967).

Apomorphine is a potent non selective agonist at the D_1 and D_2 dopamine receptors, acting both pre- and post-synaptically (Lal, 1988). A century after the first preparation of apomorphine, the total synthesis of racemic R,S-apomorphine was carried out in a multistep process by Neumeyer et al. in 1970. Soon afterwards, racemic apomorphine was resolved into 6a-(−)-R and 6a-(+)-S enantiomers by Saari et al. (1973), who established that the dopaminergic activity of the drug resides principally in the levo-rotatory 6a-(−)-R isomer.

In this report, we present a novel function of apomorphine, independent from its dopaminergic activity; we provide evidence that apomorphine has a potent cytostatic effect on Chinese hamster ovary (CHO)-K1 cells. The cytostatic effect reverses to cytotoxicity as the concentration of apomorphine increases.

Materials and methods

Materials

R(−)-Apomorphine hydrochloride was purchased from Research Biochemicals International. [³H]Thymidine was from NEN. Spiperone, NADH, sodium pyruvate and all the components for cell culture medium (except for the fetal bovine serum, which was purchased from Celbio) were from Sigma. Tissue culture supplies were from COSTAR. SCH23390 was kindly given by Dr. G. Demontis.

Cell culture

CHO-K1 cells were grown at 37°C under an atmosphere of 5% CO_2/95% air in Dulbecco's modified Eagle's medium (DMEM) supplemented with 10% (vol/vol) fetal bovine serum, 2% (vol/vol) 20 mM L-glutamine, 1% (vol/vol) penicillin (10,000 units/ml) and streptomycin (10 mg/ml) solution and 1% (vol/vol) minimal essential medium non-essential amino acid solution. Cells were seeded at a density of 10^4 cells/ml in 12-well plates (22.6-mm diameter) in a final volume of 1 ml of medium. The day after plating, the cells were exposed to treatment. After 72 hr, the number of viable cells was ascertained: cells were lifted from the wells with trypsin and counted with a hemacytometer. Cell viability was assayed by exclusion of trypan blue.

[³H]Thymidine incorporation

Cells were incubated for 72 h in the presence of apomorphine. Four hr prior to harvesting, 0.5 μCi of [³H]thymidine (NEN, 85 Ci/mmol) was added to the culture. Incorporation was stopped by washing the cells three times with ice-cold phosphate buffer saline. The cells were scraped into 0.5 ml of 0.2 N NaOH containing 0.2% Triton X-100. An aliquot of 10 μl was taken for protein determination, and 0.5 ml of HCl 0.2 N was added to neutralize the pH. The radioactivity was counted with 10 ml of INSTA-Gel (Packard) in a liquid scintillation β-counter.

Cell cycle analysis

CHO-K1 cells were plated in a $175\,cm^2$ flask (3×10^5 in 20 ml of medium). On the following day, cells were treated with apomorphine (or saline for control) at a final concentration of $10\,\mu M$ and incubated for 72 hr. On the day of the assay, cells were washed twice with PBS and lifted from the flask with trypsin. Then they were spun down at $2{,}000\,g$ for 5 min and resuspended in 5 ml PBS (this step was repeated 3 times). After the third centrifugation, cells were resuspended in PBS containing 100 mg/ml RNAse A (Boeringer) and 50 mg/ml propidium iodide (Sigma) and stained for 30 min at room temperature. Cells were analyzed with a FACSort flow cytometer (Becton-Dickinson, San Jose, CA, USA). Data analysis was performed with CELLQuest software programs (Becton-Dickinson) to determine percentages of cells in the G0/G1, S, and G2/M phases of the cell cycle.

Measurement of LDH release

Cytotoxicity was estimated by measuring the leakage of LDH. In brief, after incubating CHO-K1 cells with apomorphine at $37°C$ in a 5% CO_2 incubator for 72 hr, the media were collected and centrifuged at $800\,g$ for 5 min. The supernatant was used for the assay of LDH activity. The reaction was initiated by mixing 0.1 ml of cell-free medium with 100 mM potassium phosphate buffer (pH 7.5) containing 0.11 mM NADH and 0.7 mM sodium pyruvate in a final volume of 3 ml. The rate of absorbance decrease at 340 nm was monitored. The LDH activity was expressed as U/l.

Statistics

Data are means $\pm$ SEM of at least three experiments, each performed in triplicate. Student's two-tailed t test for unpaired data was used to evaluate the statistical differences of the means; $p < 0.05$ was accepted as being significant.

Results

Antiproliferative effect of apomorphine on CHO-K1 cell lines

The dopamine receptor agonist apomorphine was tested for its ability to suppress CHO-K1 cell proliferation. As shown in Fig. 1a (inset), apomorphine inhibited cell proliferation in a dose-dependent manner. The EC_{50} of apomorphine-induced inhibition of cell proliferation determined by cell counting was $3.24 \pm 0.07\,\mu M$. Remarkably, the dose-response curve obtained by measuring the incorporation of [³H]thymidine was practically identical to the previous one, giving an EC_{50} of $3.52 \pm 0.04\,\mu M$. A high percentage of CHO-K1 fibroblasts in the treated wells showed a rounded-shape morphology.

In order to demonstrate that the antiproliferative effect was not mediated by dopamine receptors, we initially tested CHO-K1 cells for the presence of [³H]SCH23390 and [³H]spiperone binding. We were not able to detect any binding with these two radioligands (data not shown). As these antagonists

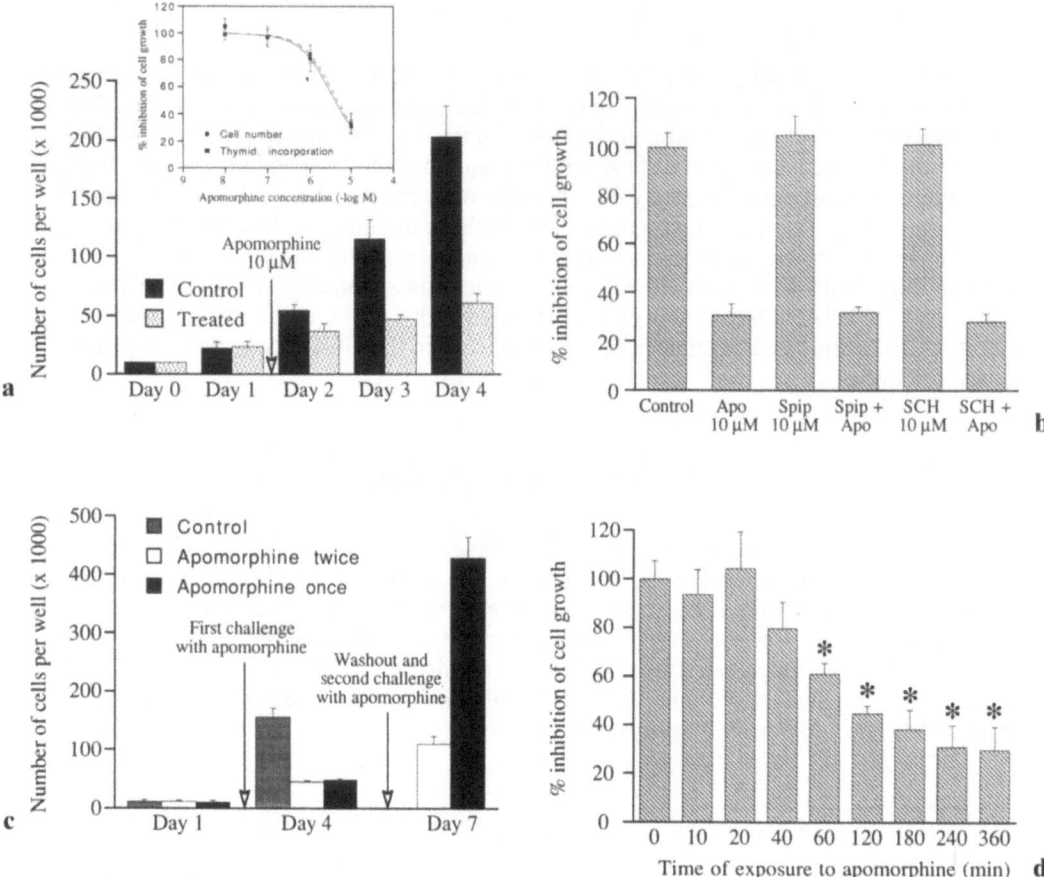

Fig. 1. Antiproliferative effect of apomorphine (10 μM) on CHO-K1 cells. **a** Time-course of apomorphine inhibition of cell growth (in the inset: dose-response curves obtained with cell-counting method and [³H]thymidine incorporation). **b** Antiproliferative effect of apomorphine in the presence of dopamine antagonists (*Apo* apomorphine; *Spip* spiperone; *SCH* SCH23390). **c** Recovery of cell growth after removal of apomorphine from the cell bath. **d** Minimal exposure time efficacy of apomorphine-induced inhibition of cell growth. *Significantly different from control: p < 0.05 (Student's t test for unpaired data)

can recognize D_1- and D_2-like receptors, the lack of binding indicates the absence of dopamine receptors. Yet this does not exclude the possibility that dopamine receptors present at a very low expression level could pass undetected by binding. Therefore, to exclude this possibility, we tested the inhibitory action of apomorphine on CHO-K1 cell growth in the presence of SCH23390 and spiperone. Figure 1b shows clearly that neither SCH23390 nor spiperone, at a concentration of 10 μM (well beyond their value of K_d for the dopamine D_1- and D_2-like receptors respectively) were able to antagonize the antiproliferative effect of apomorphine.

We also studied the time-course of apomorphine-induced inhibition of cell proliferation. As is shown in Fig. 1a, the effect was already evident after one

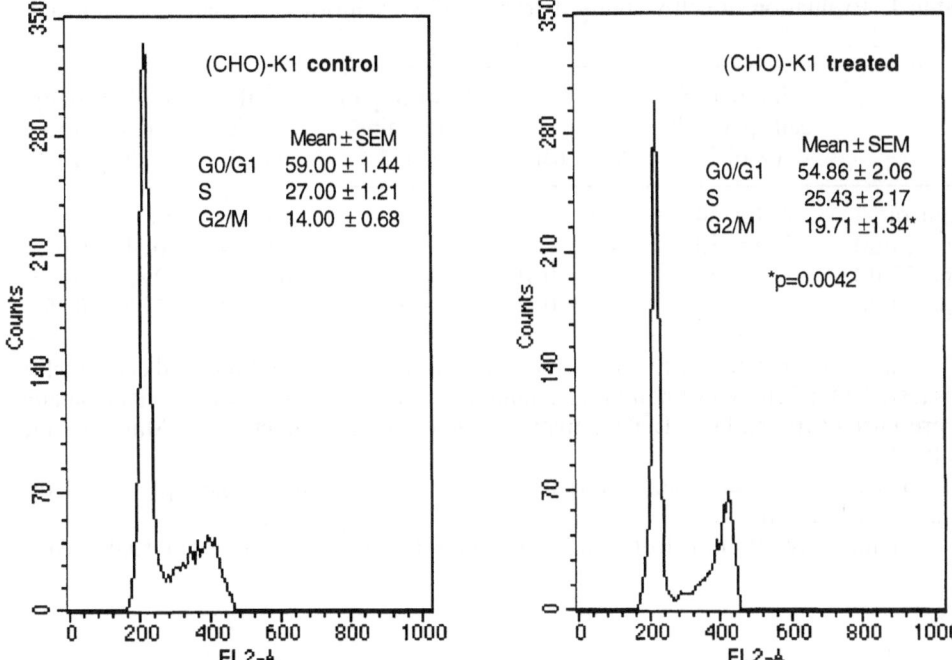

Fig. 2. Effect of apomorphine on DNA content of CHO-K1 cells after 72 hr of treatment, as determined by flow cytometry, following staining with propidium iodide. The graphs represent one of the 7 experiments, while numbers are the means ± SE of all the experiments. *Significantly different from control: p < 0.05 (Student's t test for unpaired data)

day of treatment, and increased considerably on the following two days. The antiproliferative effect of apomorphine (10 μM) was reversible, as is shown by the fact that cell growth restarted after removal of apomorphine from the cell bath on the third day (Fig. 1c). Apomorphine exerts its effect early during incubation; CHO-K1 cells exposed to apomorphine for a period as short as 1 h and then allowed to grow for three days were significantly reduced in number with respect to untreated control cells. After four hours of exposure to apomorphine 10 μM, the antiproliferative effect was similar to that seen when apomorphine was present in the bath for all three days (Fig. 1d).

Effect of apomorphine on the CHO-K1 cell cycle

To investigate the effect of apomorphine on the cell cycle of CHO-K1 cells, we studied the change in the intensity of DNA fluorescence using flow cytometric analysis of propidium-iodide (PI)-stained purified cell nuclei after 72 hr of exposure to apomorphine. In Fig. 2, the DNA histogram shows that apomorphine (10 μM) increased the number of cells in the G2/M phase (from 14.0 ± 0.68 to 19.7 ± 1.34) while it had no significant effect on the G0/G1 and S phases. The effect of apomorphine on the cell cycle was reversible after replacement of the agent with fresh medium (data not shown).

Table 1. Evaluation of cell viability determined by exclusion of trypan blue and LDH activity

	Number of cells per well ($\times 10^{-3}$)	% of cells	% of trypan blue stained cells*	LDH activity (U/l)	LDH activity/ number of cells ($\times 10^{-3}$)
Control	474 ± 54	100 ± 11	0	60 ± 6	0.126 ± 0.013
Apo 10 μM	200 ± 30	42 ± 6.3[a]	0	42 ± 7	0.21 ± 0.035
Apo 25 μM	22 ± 3	4.6 ± 0.9[a]	18	63 ± 8	2.86 ± 0.364[a]
Apo 50 μM	4 ± 0.8	0.8 ± 0.1[a]	100	91 ± 3	22.7 ± 0.75[a]

Cells were incubated with different concentrations of apomorphine and cell viability was tested after 72 hr of incubation. The number of cells was counted with hemacytometer in presence of trypan blue. LDH activity was determined as described in Materials and methods.

* The percentage exclusion of trypan blue refers to the number of cells presents in the well for each treatment.

[a] Significantly different from values obtained in the absence of apomorphine ($p < 0.05$).

Toxic effect of apomorphine on CHO-K1 cells

After three days of treatment with 10, 25 and 50 μM apomorphine, the percentage of cells present in the wells compared with controls were respectively 42, 4.6 and 0.8% (Table 1). Of these cells, respectively, 0, 18 and 100% were dead, as determined by exclusion of trypan blue. The LDH activity normalized for the number of cells rose exponentially as the concentration of apomorphine increased from 10 to 50 μM. While at 10 μM apomorphine the LDH activity was not different from controls, at 25 and 50 μM it was clearly higher (Table 1).

Discussion

Apomorphine is a drug more than one hundred years old, which has met with alternate fortunes in the symtomatic therapy of Parkinson's disease (Lees, 1993). While its therapeutic effect on Parkinson's disease is mediated by its activity as an agonist at dopamine receptors, other functions have also been attributed to this molecule (Iversen et al., 1976; Sheppard and Wiggan, 1971; Shen et al., 1984). In this article, we report an additional property of this compound: apomorphine has an antiproliferative effect on CHO-K1 fibroblasts, with an EC_{50} in the low micromolar range. Our data showed clearly that this effect was not due to activation of dopamine receptors, as there are no dopamine receptors in these cells. The antiproliferative effect of apomorphine was observed up to 10 μM, and at this concentration no significant cell death was detectable. Higher concentrations, instead, resulted in a progressive toxicity of apomorphine for the CHO-K1 cells, with a complete destruction of the colony at $50\,\mu M$. As reported in the Results section, a high percentage of cells showed a round-shaped morphology; this alteration was reversible with the suspension of apomorphine treatment on the third day (see below). Furthermore, it was often noted that solitary cells grew much more slowly than aggregated cells, leading to a non-homogeneous distribution on the third day of incubation. This could be explained by the secretion of autocrine growth factors by CHO-K1 cells into the surrounding medium, which would support cell proliferation.

The antiproliferative effect of apomorphine was reversed by suspension of the treatment on the third day of incubation with a restoration of the polygonal morphology of the cells and the growth rate. It is interesting to note that the reduction in the number of cells exposed for 4hr to apomorphine was similar to that observed when apomorphine was present in the bath for all three days. This indicates that apomorphine exerts its antiproliferative activity early during incubation and the effect lasts independently of its presence in the incubation bath.

This can be explained by the fact that apomorphine undergoes a rapid autoxidation at neutral pH with the formation of a melanin-like polymer as an end product (Gassen et al., 1996). In an aqueous solution, 90% of apomorphine is consumed after 2hr. Since the unoxidized form is probably responsible for priming the antiproliferative effect, apomorphine should exert its activity only during the first few hours. In this context, it is clear that the permanence of apomorphine in the bath for all three days is uninfluential. The long-lasting effect could be explained by a permanent alteration of some protein components of the cell cycle which should have been resynthesized in order to restore cell growth.

In our experiment apomorphine induced an accumulation of CHO-K1 fibroblasts in the G2/M phase. This effect is common to some anticancer agents like quercetin (Kang and Liang, 1997) and genistein (Matsukawa et al., 1993). Quercetin in particular has in common with apomorphine a catechol ring in its chemical structure. It has been suggested that the antiproliferative effect of quercetin could be mediated in part by inhibition of protein Kinase C (PKC) (Ferriola et al., 1989; Kang and Liang, 1997), as there is considerable evidence that PKC is implicated in the growth and proliferation of cells and development of cancer (Nishizuka, 1989). Remarkably, apomorphine has recently been found to inhibit PKC (Wang et al., 1997). It is worth noting that the IC_{50} value found to inhibit PKC ($8\,\mu M$) is similar to the EC_{50} we calculated for the antiproliferative effect of apomorphine on CHO-K1 cells.

As mentioned above, apomorphine has a catechol ring in its structure. Compounds with a catechol structure have metal-chelating properties and can act as reducing agents (Liu and Mori, 1993; Gassen et al., 1998). Therefore apomorphine may, on the one hand, inhibit metal-catalyzed free radical processes and act as a radical scavenger, but on the other hand, as a reducing agent it may also contribute to the generation of highly toxic hydroxyl radicals (HO·) by maintaining iron in the ferrous state. HO· radicals are most probably responsible for most of the damage caused by free radicals in vivo (Ben-

Shachar et al., 1991). In addition, catechol can undergo the formation of quinone intermediates (Graham, 1978; Rosenberg, 1988). These unstable, electron-deficient compounds can form catechol conjugates by reacting with cellular nucleophiles such as the sulfhydryl groups of cysteines, which are present as components of proteins (Hastings and Zigmond, 1994). It is likely that the overall configuration of apomorphine can target the molecule to some preferential site where it can bind and exert its toxic effect, either by generating hydroxyl radicals or by forming catechol conjugates.

References

Anden NE, Rubenson A, Fuxe K, Hokfelt T (1967) Evidence for dopamine receptor stimulation by apomorphine. J Pharm Pharmac 19: 627–629

Ben-Shachar D, Riederer P, Youdim MBH (1991) Iron-melanin interaction and lipid peroxidation: implication for Parkinson's disease. J Neurochem 57: 1609–1614

Ernst AM (1967) Mode of action of apomorphine and dexamphetamine on gnawing compulsion in rats. Psychopharmacologia 10: 316–323

Ferriola PC, Cody V, Middleton EJr (1989) Protein kinase C inhibition by plant flavonoids. Kinetic mechanisms and structure-activity relationships. Biochem Pharmacol 38: 1617–1624

Gassen M, Glinka Y, Pinchasi B, Youdim MBH (1996) Apomorphine is a highly potent free radical scavenger in rat brain mitochondrial fraction. Eur J Pharmacol 308: 219–225

Gassen M, Grass A, Youdim MBH (1998) Apomorphine enantiomers protect cultured pheochromocytoma (PC12) cells from oxidative stress induced by H_2O_2 and 6-hydroxydopamine. Mov Disord 13: 242–248

Graham DG (1978) Oxidative pathways for catecholamines in the genesis of neuromelanin and cytotoxic quinones. Mol Pharmacol 14: 633–643

Hastings TG, Zigmond MJ (1994) Identification of catechol-protein conjugates in neostriatal slices incubated with [³H]dopamine: impact of ascorbic acid and glutathione. J Neurochem 63: 1126–1132

Iversen LL, Rogawski MA, Miller RJ (1976) Comparison of the effects of neuroleptic drugs on pre- and postsynaptic dopaminergic mechanism in the rat striatum. Mol Pharmacol 12: 251–262

Kang TB, Liang NC (1997) Studies on the inhibitory effects of quercetin on the growth of HL-60 leukemia cells. Biochem Pharmacol 54: 1013–1018

Lal S (1988) Apomorphine in the evaluation of dopaminergic function in man. Prog Neuropsychopharmacol Biol Psychiatry 12: 117–164

Lees AJ (1993) Dopamine agonists in Parkinson's disease: a look at apomorphine. Fundam Clin Pharmacol 7: 121–128

Liu J, Mori A (1993) Monoamine metabolism provides an antioxidant defense in the brain against oxidant- and free radical-induced damage. Arch Biochem Biophys 302: 118–127

Matsukawa Y, Marui N, Sakai T, Satomi Y, Yoshida M, Matsumoto K, Nishino H, Aoike A (1993) Genistein arrests cell cycle progression at G2-M. Cancer Res 53: 1329–1331

Neumeyer JL, Neustadt BR, Weinhardt KK (1970) Aporphines. V. Total synthesis of (plus or minus)-apomorphine. J Pharm Sci 59: 1850–1852

Nishizuka Y (1989) Studies and prospectives of the protein kinase C family for cellular regulation. Cancer 63: 1892–1903

Rosenberg PA (1988) Catecholamine toxicity in cerebral cortex in dissociated cell culture. J Neurosci 8: 2887–2894

Saari WS, King KK, Lotti KK (1973) Synthesis and biological activity of (6aS)-10,11-dihydroxyaporphine, the optical antipode of apomorphine. J Med Chem 16: 171–172

Shen RS, Smith KK, Davis PJ, Abell CW (1984) Inhibition of dihydropteridine reductase from human liver and rat striatal synaptosomes by apomorphine and its analogues. J Biol Chem 259: 8994–9000

Sheppard H, Wiggan G (1971) Different sensitivities of the phosphodiesterases (adenosine-3′,5′-cyclic phosphate 3′-phosphohydrolase) of dog cerebral cortex and erythrocytes to inhibition by synthetic agents and cold. Biochem Pharmacol 20: 2128–2130

Wang BH, Lu ZX, Polya GM (1997) Inhibition of eukaryote protein kinases by isoquinoline and oxazine alkaloids. Planta Med 63(6): 494–498

Authors' address: Prof. Giovanni U. Corsini, Dipartimento di Neuroscienze, Sezione di Farmacologia, Università di Pisa, Via Roma 55, 1-56100 Pisa, Italy

Potent neuroprotective and antioxidant activity of apomorphine in MPTP and 6-hydroxydopamine induced neurotoxicity

E. Grünblatt[1], S. Mandel[1], M. Gassen[2], and M. B. H. Youdim[1,3]

[1] Technion — Faculty of Medicine, Eve Topf and U.S. National Parkinson's Foundation Centers for Neurodegenerative Diseases, Bruce Rappaport Family Research Institute and Department of Pharmacology, Haifa, Israel
[2] Merck KGaA, Biomedical Research CNS, Darmstadt, Federal Republic of Germany
[3] N.I.H. Fogarty International Center and NIMH, Bethesda, MD, U.S.A.

Summary. Apomorphine is a potent radical scavenger and iron chelator. In vitro apomorphine acts as a potent iron chelator and radical scavenger with IC_{50} of $0.3\,\mu M$ for iron ($2.5\,\mu M$) induced lipid peroxidation in rat brain mitochondrial preparation, and it inhibits mice striatal MAO-A and MAO-B activities with IC_{50} values of $93\,\mu M$ and $241\,\mu M$. Apomorphine ($1-10\,\mu M$) protects rat pheochromocytoma (PC12) cells from 6-hydroxydopamine ($150\,\mu M$) and H_2O_2 ($0.6\,mM$) induced cytotoxicity and cell death. The neuroprotective property of (R)-apomorphine, a dopamine D_1-D_2 receptor agonist, has been studied in the MPTP (N-methyl-4-phenyl-1,2,3,6-tetrahydropyridine) model of Parkinson's disease. (R)-apomorphine (5–10 mg/kg, s.c.) pretreatment in C57BL mice, protects against MPTP (24 mg/kg, I.P) induced loss of nigro-striatal dopamine neurons, as indicated by striatal dopamine content, tyrosine hydroxylase content and tyrosine hydroxylase activity. It is suggested that the neuroprotective effect of (R)-apomorphine against MPTP neurotoxicity derives from its radical scavenging and MAO inhibitory actions and not from its agonistic activity, since the mechanism of MPTP dopaminergic neurotoxicity involves the generation of oxygen radical species induced-oxidative stress.

Introduction

Although the etiology of Parkinson's disease (PD) is still not known, a significant body of evidence from our laboratory, as supported by others, points to the presence of ongoing oxidative stress (OS), selectively in the substantia nigra pars compacta (SN) of Parkinsonian brains (Youdim et al., 1993; Gerlach et al., 1994; Jenner et al., 1996; Olanow et al., 1996). The evidence for OS includes proliferation of reactive microglia with highly significant increase in cytokine TNF-alpha, interleukin 1 and interleukin 6, an increase in the levels of iron, BCP2, lipofucin, in monoamine oxidase B activity and in lipid peroxidation and a reduction in calcium binding protein (e.g. calbindin, D 28), in glutathione (GSH) and in mitochondrial complex I activity (Youdim et al., 1993; Jenner et al., 1996). Increase of iron in the SN of idiopathic PD, N-methyl-4-phenyl-1,2,3,6- tetrahydropyridine (MPTP) and 6-hydroxydopamine (6-OHDA) induced Parkinsonism, could be a prominent feature in the activation of microglia and the increase of cytokines observed. Recent studies have shown that increased iron in macrophages and microglia, as seen in PD (Jellinger et al., 1990), is linked to iron induced gene regulation of interleukin 1, interleukin 6 and TNF-alpha (Rogers et al., 1990; Seiser et al., 1993; Simeonova et al., 1995). These iron and cytokines could act in concert to initiate oxidative stress via promotion of oxygen radical species such as super-oxide (O_2°), hydroxyl radicals (OH°) and nitric oxide (NO) (Youdim et al., 1990; Adamson et al., 1993).

Indeed, in the animals models of PD, both MPTP and 6-OHDA produce almost exactly similar biochemical changes, presumably due to their ability to cause OS (Cohen et al., 1994). In these models, antioxidants like vitamin E, iron chelators (e.g. desferal) (Ben-Shachar and Youdim, 1991; Ben-Shachar et al., 1991; Sengstock et al., 1994), monoamine oxidase (MAO) B (Heikkila et al., 1984) and nitric oxide synthase inhibitors (Matthews et al., 1997) induce neuroprotection. In line with this concept there has been a concerted attempt to look for and develop potential neuroprotective drugs with antioxidant and iron chelator activity. However some drugs either do not cross the blood brain barrier or are ineffective and/or toxic.

(R)-Apomorphine (APO), the dopamine D_1–D_2 receptor agonist, is one of the most potent anti Parkinson drug available and is an effective replacement of L-dopa (3,4-dihydroxyphenylacetic acid) in the therapy of late stage PD. However, its rapid metabolism and pharmacokinetics may be a limiting factor in therapy. This disadvantage has been overcome by the use of a continuous-self injector system (Gancher et al., 1995). The recent clinical studies have indicated that long term treatment with APO can result in weaning off the patient from L-dopa (Less, 1993; Colzi et al., 1998). These results have been interpreted as indicating either the necessity of continuous D_1–D_2 dopamine receptor stimulation and possible neuropro-tection, in order to achieve a better therapeutic response (Gassen et al., 1996; Gassen et al., 1998a). Compounds with a catechol structure have metal chelat-ing properties and can act as reducing agent (Liu et al., 1993), therefore, APO can inhibit metal-catalyzed free radical processes and act as a radical scavenger (Gassen et al., 1996). As a reducing agent, APO can also contribute to the generation of highly toxic OH° by maintaining iron in the ferrous state. The overall manner by which antioxidant drug affects the level of OS depends on the balance between radical scavenging and radical activating properties.

Our previous studies have shown that APO is highly potent radical scav-enger and iron chelator (Gassen et al., 1996; Gassen et al., 1998b), displaying IC_{50} value of 0.3–0.6 μM. These properties are directly linked to its ability to inhibit brain mitochondrial lipid peroxidation and protein oxidation (40%) (Gassen et al., 1996). Another evidence to its protective action was shown against the cytotoxic action of hydrogen peroxide (H_2O_2) and 6-OHDA in PC12 cells (Gassen et al., 1998a).

The aim of our study was to investigate whether APO will be neuroprotective in animal models of PD with MPTP and 6-OHDA. This was considered logical since these neurotoxins are thought to produce their dopaminergic neurotoxicity via generation of oxygen radical species (Ebadi et al., 1996), liberation of iron in the substantia nigra pars compacta (Mochizuki et al., 1994; Temlett et al., 1994), depletion of reduced glutathione (GSH) (Di Monte et al., 1987) and inhibition of mitochondrial complex I (Singer et al., 1987; Glinka and Youdim, 1995; Glinka et al., 1996). Furthermore, iron chelators and radical scavengers are able to protect not only against MPTP-induced neurotoxicity (Lan et al., 1997; Santiago et al., 1997) but also against that induced by 6-OHDA and hydrogen peroxide (Cadet et al., 1989; Ben-Shachar and Youdim, 1991; Perumal et al., 1992; Glinka et al., 1996; Glinka et al., 1997; Gassen et al., 1998).

Free radical scavenging property of apomorphine

The protecting effect of APO was examined in lipid peroxidation and protein carbonyl formation after ascorbate/iron-induced free radical formation in rat brain mitochondrial fractions. Addition of submolecular concentration of APO, to the ascorbate/iron-induced lipid peroxidation in rat brain mitochondrial fraction, induced a marked reduction in the formation of thiobarbituric acid reactive substances (a substance marking the extend of lipid peroxidation) as compared with samples containing only ascorbate and iron. The effectiveness of APO inhibition was dependent on the iron concentration; the IC_{50} varied in the range between 0.1 and 1 µM. This influence could be indirect as the overall intensity of free radical formation is also subject to changes in iron concentration. The exact values obtained: IC_{50} was 0.3 µM for 2.5 µM and 0.6 µM for 5 µM iron (Table 1). The sigmoid character of the concentration-response relation for the inhibition was stronger at the higher iron concentration, and the apparent Hill coefficient rose from 1.7 (2.5 µM $FeSO_4$) to 4.2 (5 µM $FeSO_4$) (Table 1). This is consistent with a multi-step oxidation of APO with a melanin-like polymer as an end product.

For a more solid basis to the assumption that APO is a free radical scavenger, a time course of APO oxidation was monitored. During oxidation an intensive green chromophore (λ_{max} = 619 nm) is formed which has been used to monitor the reaction, as there is little interference with other components in the assay. The color formation reflects a complicated multistep process with autoxidation of APO itself being only the initial step. The reaction is considerably slowed down by 50 µM ascorbate, but addition of 5 µM iron give no significant change. In a system containing brain mitochondria, APO is oxidized even in the presence of ascorbate/iron. This reaction reflects the free radical scavenging effect of APO, which leads to decrease in formation of thiobarbituric acid reactive substances.

Table 1. Inhibition of ascorbate/iron induced lipid peroxidation by apomorphine, dopamine, and desferrioxamine

FeSO$_4$ (μM)[a]	Apomorphine		Dopamine	Desferrioxamine
	2.5	5.0	2.5	2.5
IC$_{50}$ [μM][b]	0.28 ± 0.02	0.61 ± 0.02	6.59 ± 0.2	0.78 ± 0.04
Max inhibition[c] [%]	92 ± 1	93 ± 2	93 ± 1	75 ± 1
Hill coefficient[d]	1.7 ± 0.1	4 ± 0.3	1.0 ± 0.1	0.9 ± 0.15

[a] Concentration of ascorbate was 50 μM in all cases. [b] Obtained from regression data (mean ± SE, n = 6). [c] Maximum inhibition as determined from a triplicate experiment (mean ± SEM). [d] Apparent values (mean ± SE, n = 5), as obtained from the slope of the cooperativity plot (Hill plot), log [apomorphine] vs. log (I/I$_{max}$ − I))

APO was compared with dopamine (DA) and desferrioxamine, a potent iron chelator that is able to inhibit iron catalyzed lipid peroxidation (Sorrenti et al., 1994). It was found that DA inhibited the formation of thiobarbituric acid reactive substances in a similar manner as APO, while the effective concentration being twenty times higher (Table 1). Desferrioxamine was not able to block lipid peroxidation completely. At maximum inhibition, 16% of the activity remained (Table 1). On the basis of this data, iron chelation by dopamine may be a major contribution to the observed inhibition of TBARS formation. This can be ruled out for apomorphine, as apomorphine provides complete inhibition at concentrations much lower than the Fe^{2+}-concentration.

Apomorphine also protects against oxidation of proteins. Measuring protein carbonyl formation is a more specific but less sensitive method to assay free radical damage in biological system. Under conditions of OS, proline, arginine, lysine and threonine residues are converted into aldehydes and ketones (Stadtman, 1993), which can be labeled with specific reagents, like 1,4-dinitrophenylhydrazine. Addition of 100 μM APO, in protein oxidation assay with 250 μM ferrous iron and 15 mM ascorbic acid, gave a significant decrease of protein oxidation (40%).

Protection from oxidative stress in PC12 cells by apomorphine

A key question that remained after these results concerns the balance between catecholamine toxicity and possible beneficial effects due to antioxidation. We looked at this problem in PC12 cell culture, a well established system to study apoptotic and necrotic cell death (Vimard et al., 1996). OS can be induced by various agents like H$_2$O$_2$, organic hydroperoxides, or 6-OHDA.

We treated PC12 cells with H$_2$O$_2$ and 6-hydroxydopamine and observed cell death in a concentration dependent manner within 24 h. There was no

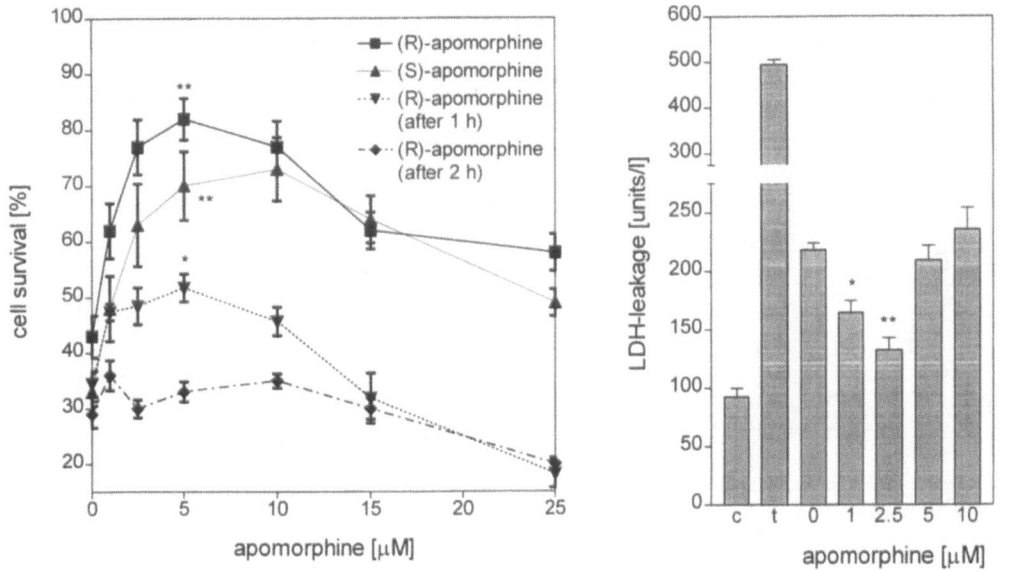

Fig. 1. H_2O_2 toxicity in PC12 cell culture and protection by apomorphine. **a** Cells were treated with 0.6 mM H_2O_2 and apomorphine. Cell viability was assayed with MTT 24 h later and expressed as percent of controls (Data ± S.E.M., n = 8). The difference between (R)- and (S)-apomorphine is not significant (two-way-ANOVA: p = 0.08) **b** LDH leakage as determined 24 h after treatment with H_2O_2 (0.6 mM) and apomorphine, (c controls t total activity after lysis of the cells, Data ± S.E.M., n = 5). Peak data points were compared with controls by Mann-Whitney U-test (*: p < 0.05; **: p < 0.01)

significant difference of the sensitivity between cells that were grown in medium containing 15% serum (1/3 fetal calf serum, 2/3 horse serum) and those that had been differentiated for six days with additional 100 µg/ml 7S-NGF (Greene et al., 1987), if all the nerve growth factor (NGF) had been washed out prior to the experiment. Although it takes 24 h to observe the maximum cell death, only two hours exposure to the toxic agent is sufficient to induce the full damage. The viability of the cells has been tested by measuring the conversion of 3-(4,5-dimethylthiozol-2-yl)-2,5-diphenyltetrazolium bromide (MTT) into a colored formazane derivative and lactate dehydrogenase (LDH) leakage.

Exact EC_{50} values were obtained: 400 µM for H_2O_2 and 150 µM for 6-OHDA were necessary to kill 50% of the cultured cells. In this system, APO was tested for its ability to protect PC12 cells from the oxidative insults. At the same time, the toxicity can be monitored to obtain information about the therapeutic window of the agent. We found that APO is by far more efficient as an antioxidant: only 5 µM improve the rate of survival from 50% to 85% in the presence of 400 µM H_2O_2 (Fig. 1a). Both (R)-APO, which is active as dopaminergic agonist, and the inactive (S)-enantiomer showed protective properties against H_2O_2. Any protection against H_2O_2 by APO depended on the presence of the drug during the insult. Preincubation with APO and washout prior to H_2O_2 addition or addition of H_2O_2 one hour after the toxin

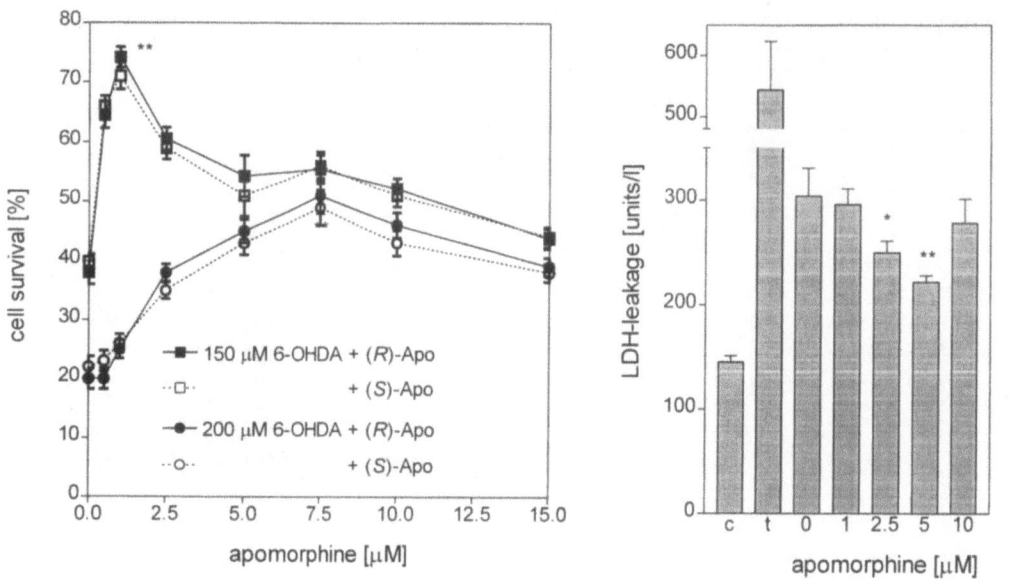

Fig. 2. 6-Hydroxydopamine toxicity in PC12 cell culture and protection by apomorphine.
a Cells were treated with 6-hydroxydopamine and apomorphine. Cell viability was assayed with MTT 24 h later and expressed as percent of controls (Data ± S.E.M., n = 8).
b LDH leakage as determined 24 h after treatment with 6-hydroxydopamine (150 μM)
and apomorphine (Data ± S.E.M., n = 5). Peak data points were compared with controls
by Mann-Whitney U-test (*: p < 0.05; **: p < 0.01)

did not improve the survival as compared with controls only treated with the
oxidant (Fig. 1a). With concentrations exceeding 10 μM, the toxicity of APO
became increasingly dominant. APO alone induced death of PC12 cells with
EC_{50} = 100 μM; this concentration caused an almost complete cell loss in the
presence of H_2O_2.

APO was able to provide protection against 6-OHDA insults. The survival
rate after 150 μM 6-OHDA (EC_{50}) was improved to 70% with only 1 μM
apomorphine (Fig. 2a). This is the first example a catecholamine to attenuate
the toxicity of 6-OHDA in cell culture.

The protective effect of APO could also be demonstrated by measuring
the LDH leakage induced by H_2O_2 (0.6 mM) and 6-OHDA (150 μM). In the
presence of APO (1–10 μM), the effects of both toxins were markedly attenuated. Apart from slight variations, the LDH data correlates with the findings
in the MTT assay (Fig. 1b and 2b).

Neuroprotection activity of apomorphine against MPTP induced neurotoxicity in mice

The possible neuroprotective effect of APO was examined in-vivo by analyzing the striatal content of DA and its metabolites (3,4-dihydroxyphenylacetic
acid (DOPAC), homovanillic acid (HVA)) by HPLC. MPTP treatment

Table 2. Striatal DA and metabolites content in C57BL mice treated with APO and MPTP

Treatment	DA [pmol/mg tissue]	DOPAC [pmol/mg tissue]	HVA [pmol/mg tissue]	DOPAC + HVA / DA	HVA / DA
Control	47.2 ± 3.7	2.28 ± 0.19	21.6 ± 2.1	0.50 ± 0.03	0.48 ± 0.05
APO (10 mg/kg/d)	45.6 ± 5.9	2.58 ± 0.20	14.3 ± 2.1	0.40 ± 0.04	0.31 ± 0.04
MPTP (24 mg/kg/d)	11.1 ± 1.2***a	0.85 ± 0.11***a	22.4 ± 0.7	1.73 ± 0.16	1.78 ± 0.17**a
APO + MPTP	34.0 ± 5.2**b	2.05 ± 0.14***b	14.9 ± 4.4	0.29 ± 0.02*a	0.24 ± 0.02*a**b

Mice (8–10 per treatment group) 1 were injected once daily with APO (10 mg/kg) followed immediately by a dose of MPTP (24 mg/kg), for 5 days and sacrificed 2 days later. Controls received saline only. DA and metabolites were analyzed by HPLC

*$p < 0.01$; **$p < 0.005$; ***$p < 0.001$; [a], vs. control, [b], vs. MPTP

(24 mg/kg/day/5 days, sacrifice 2 days from last injection) caused significant reductions in DA and DOPAC concentrations (~65% and ~60%, respectively) as predicted (Fig. 3a and 3b, Table 2). However pretreatment of mice with APO followed by MPTP resulted in partial protection of neurodegeneration as indicated by DA and DOPAC levels. Starting in the dose of 5 mg/kg/day/5 days APO there is a significant elevation to the control levels (DA to 50% and DOPAC to 60% of control), whereas the 10 mg/kg/day/5 days APO dose almost completely restored the levels of DA and DOPAC to control values (72% and 90% of control, respectively), (Fig. 3a and 3b). The ratio of (DOPAC + HVA)/DA and HVA/DA were significantly increased in MPTP treated mice (350% and 370%, respectively) while the combination of APO (10 mg/kg) followed by MPTP resulted in reduced ratios (40% and 50%, respectively). Treatment of APO alone had no significant effect on striatal DA nor DOPAC and caused only a slight reduction in HVA and in the turnover, Table 2.

The possible inhibitory effect of APO on striatal MAO-A and MAO-B (the enzyme responsible for metabolizing DA) activities were examined in vitro using 10 increasing concentrations of APO (0–250 μM). APO caused a dose-dependent inhibition of MAO-A and MAO-B activities with IC_{50} values of 93 μM and 241 μM, respectively.

In western blot analysis for striatal tyrosine hydroxylase (TH) content (the rate limiting enzyme for DA), MPTP treatment markedly decreased TH levels (40% of control) while pretreatment with APO (10 mg/kg/day/5 days) significantly prevented the neuronal damage (70% of control) induced by MPTP. No significant effect on TH was observed with APO alone. In addition to that, TH activity, measured in the technique developed by Kato (1981), decrease significantly by MPTP treatment of the mice (60%), while pretreatment with APO (10 mg/kg/day/5 days) prevented the decrease of MPTP significantly back to 88% TH activity (Fig. 4).

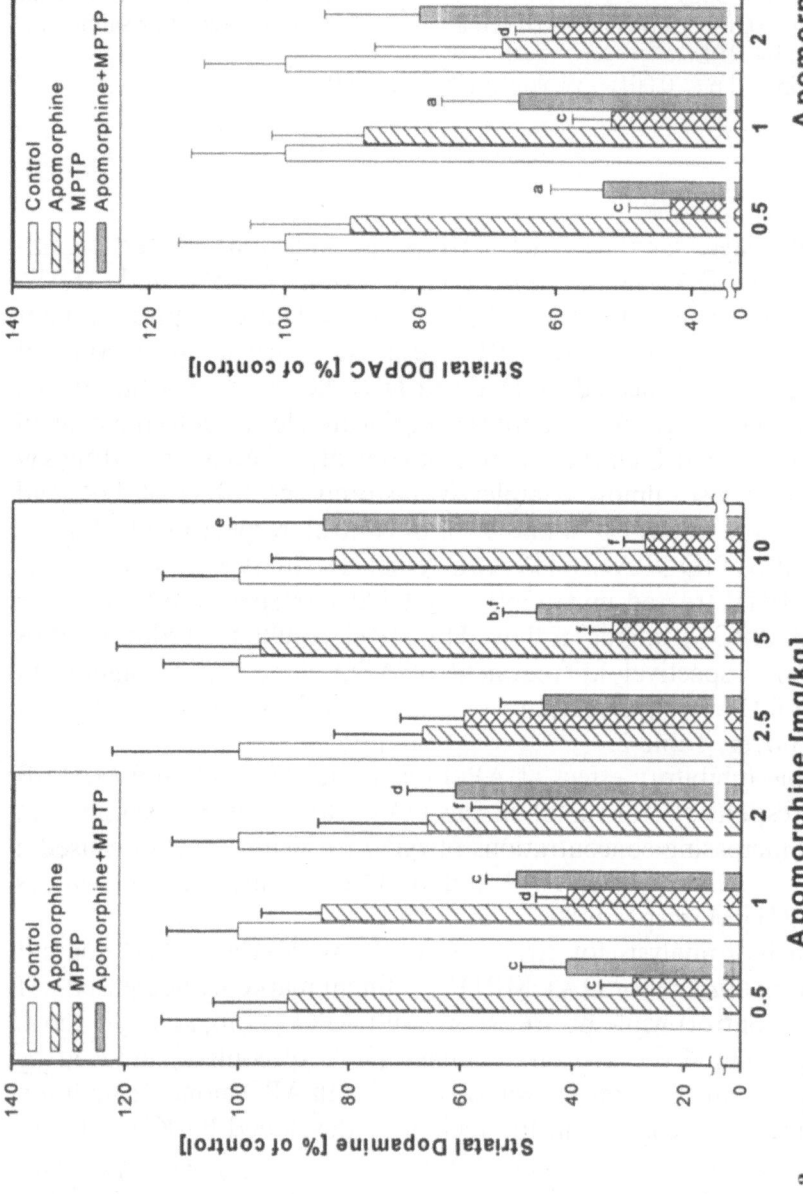

Fig. 3. Effect of apomorphine on striatal DA (**a**) and DOPAC (**b**) content. C57-BL mice were injected with apomorphine (0.5–10 mg/kg/day/5 days) followed by a dose of MPTP (24 mg/kg/day/5 days). Controls received saline or apomorphine only. Striatal DA and DOPAC were measured by HPLC. The results represent the mean ± SEM (each group 8–10 mice). *a, b* p < 0.05 (vs. control/MPTP, respectably); *c* p < 0.01 (vs. control/MPTP, respectably); *d, e* p < 0.005 (vs. control/MPTP, respectably); *f, g* p < 0.001 (vs. control/MPTP, respectably)

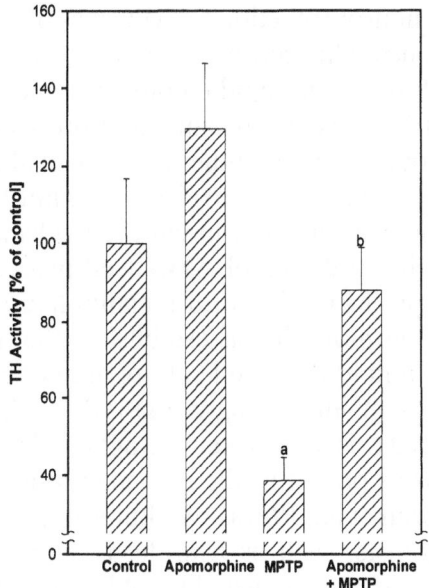

Fig. 4. TH activity in apomorphine (10 mg/kg/day/5 days) treated C57-BL mice. Mice were treated similarly to Fig. 3 and striata were dissected. The TH activity was measured according to Kato (1981) method. The results represent the mean ± SEM (each group 8–10 mice). *a* p < 0.05 (vs. control); *b* p < 0.005 (vs. MPTP)

Discussion

The different biochemical approaches to neuroprotection in Parkinson's disease reflect the current concepts of the ethiology of the disease. Antioxidant strategies, aiming at inactivating free radicals or inhibiting their formation, have been in the focus of attention: Inhibitors of MAO (e.g. L-deprenyl) (LeWitt, 1994; Przuntek, 1994) interrupt the metabolic formation of H_2O_2, while iron chelators, like desferrioxamine, block the formation of OH° radicals via the Fenton reaction. The protective effect of lazaroids and 21-aminosteroids has been originally attributed to iron chelation and prevention of membrane lipid peroxidation (Braughler et al., 1987). The current literature states potent free radical scavenging activity for this class of drugs (Zhao et al., 1995). Inhibitors of NO synthase prevent the formation of NO, which combines with $O_2^{\circ-}$, produces toxic peroxynitrite ($ONOO^-$). $ONOO^-$ is unstable and its decomposition gives rise to OH° and NO_2 radicals (Olanow, 1993). Classical free radical scavengers (e.g. ascorbate, α-tocopherol) react easily with reactive oxygen species and, thus, protect biological structures from oxidation. However, promising results with vitamin E in vitro could not be reproduced clinically to slow down the progression of Parkinson's disease (LeWitt, 1994).

We have recently shown that APO, a dopamine D_1–D_2 receptor agonist, is highly potent in protecting against MPTP and 6-OHDA neurotoxicity in vitro and in vivo. This protection is expressed by preventing the MPTP-induced loss of striatal content of DA, DOPAC, and of TH content and activity in mice. Almost complete neuroprotection by APO was observed at doses above 5 mg/kg, but not lower ones. This may be explained by the catechol containing structure of APO which results in rapid metabolism and autooxidation, since catecholamines can act both as antioxidants and pro-oxidant, depending on their concentration. In our previous studies (Gassen et al., 1996; Gassen et al., 1998a) we demonstrated that it is the oxidized and not the reduced form of APO, which initiates neuroprotection in cultured cells against H_2O_2 and 6-OHDA toxicity, while the oxidation of APO itself proceeded at much accelerated rate leading to the formation of a melanin-like polymer in complex multistep reaction. Furthermore, it was established that the neuroprotective actions of APO in vitro and in PC12 cell culture are not related to its dopamine agonistic activity. Most interestingly, (S)-APO, which is inactive as a dopamine receptor agonist, is still protective against H_2O_2 and 6-OHDA in PC12 cell culture and is as potent as the (R)-enantiomer (Sam et al., 1995a; Gassen et al., 1996; Gassen et al., 1998a). We know of no other anti-parkinson dopamine receptor agonists, such as lisuride, bromocriptine or pergolide, which show the anti-oxidant potency like APO or can exert the same neuroprotective action against H_2O_2, 6-OHDA and MPTP, either in cell culture or in vivo.

The relatively high doses of APO used in this study to achieve neuroprotection relates to the pharmacokinetic unstable and readily oxidizable nature of APO (Sam et al., 1995; Gancher, 1995). The neuroprotective effect achieved by 5 and 10 mg/kg of APO against the loss of neuronal TH is reflected by the increased levels of DA. This increase could result either from an increase in DA synthesis, caused by an elevation in the activity of TH, the rate-limiting enzyme, or from a decrease in the turnover of DA. We recently observed that APO indeed caused a 88% elevation in TH activity in MPTP treated mice, as a result of restored TH protein levels by APO. APO + MPTP treated striatal DA levels, were similar to those of the saline-treated group of mice, in spite of partial restoration of TH levels by APO treatment. This could be explained by the compensating mechanism proposed by Zigmond (1989), who demonstrated that DA levels are kept normal, so long as at least 30% of the nigro-striatal dopamine neurons remain functional. Furthermore, part of this compensating mechanism may involve the ability of APO to inhibit MAO-A and B, as observed in the in-vitro determinations. This effect may also contribute to the normalization of DA, since DA is equally metabolized by both enzymes (O'Carrol et al., 1983).

The mechanism of action of APO is best explained by its antioxidant-radical scavenging and iron chelating properties, since both enantiomers share the same properties (Sam and Verbeke, 1995; Gassen et al., 1996; Gassen et al., 1998b; Gassen and Youdim, 1998). Both 6-OHDA- and MPTP-induced neurotoxicity in-vivo result in highly elevated contents of iron in the subsatntia nigra pars compacta (Monteiro et al., 1989; Mochizuki et al., 1994; Temlett et al., 1994; Oestreicher et al., 1994). It was shown that in the case of neurotoxicity induced by 6-OHDA (Ben-Shachar et al., 1991) and MPTP (Lan et al., 1997; Santiago et al., 1997) pretreatment with the iron chelator, desferroxamine, is highly effective in neuroprotection. The mechanism of neurotoxicity is thought to be related to the ability of 6-OHDA and MPTP to release iron from ferritin and that the iron participates in the Fenton chemistry (Monteiro et al., 1989; Linert et al., 1996), resulting in generation of the highly reactive OH° radical (Ben-Shachar et al., 1991; Ben-Shachar and Youdim, 1991). It is considered that this radical propagates the process of membrane lipid peroxidation, that leads to depletion of striatal reduced GSH, which in turn, results in OS-initiated neurodegenaration (Gassen and Youdim, 1998).

We consider that APO is a unique anti-parkinsonian drug, with a number of pharmacological features that make it superior to other so called neuroprotective drugs, such as MAO-B inhibitors, vitamin E, iron-chelators and nitric oxide synthase inhibitors, which can protect against either MPTP and 6-OHDA or hydrogen peroxide, but not all three. In contrast APO has the ability to protect against the neurotoxicity of MPTP, as shown in this study, as well as against that induced by 6-OHDA, hydrogen peroxide and iron neurotoxicity (Gassen et al., 1996; Gassen et al., 1998a and b).

The superiority of APO pharmacological actions is based on:

a) its radical scavenging and iron chelating properties,
b) its ability to protect against hydrogen peroxide, 6-OHDA and iron-induced neurotoxicity in PC12 cell culture,
c) its ability to protect against MPTP-induced neurotoxicity in-vivo, in mice,
d) its ability to inhibit mitochondrial iron-induced lipid peroxidation and protein oxidation,
e) its ability to prevent 6-OHDA-induced inhibition of mitochondrial complex I activity (Gassen et al., 1996) and
f) its ability to inhibit MAO-A and B (Grünblatt et al., 1998, unpublished data).

Acknowledgement

The support of the National Parkinson Foundation (U.S.A.), the Golding Parkinson Research Fund (Technion, Haifa, Israel), and the Israel Ministry of Arts and Science are gratefully acknowledged. This paper was written while Moussa B. H. Youdim was a Scholar-in-Residence at the Fogarty International Center for Advanced Study in Health Sciences, National Institutes of Health, Bethesda, MD, U.S.A.

References

Adamson GM, Billings RE (1993) Cytokine toxicity and induction of NO synthesis activity in cultured mouse hepatocytes. Toxicol Appl Pharmacol 119: 100–107

Ben-Shachar D, Youdim MBH (1991) Intranigral iron injection induces behavioral and biochemical "parkinsonism" in rats. J Neurochem 57: 2133–2135

Ben-Shachar D, Eshel G, Finberg JPM, Youdim MBH (1991) The iron chelator desferrioxamine (desferal) retards 6-hydroxydopamine-induced degeneration of nigrostriatal neurons. J Neurochem 56: 1441–1444

Braughler JM, Pregenzer JF, Chase RL, Duncan LA, Jacobson EJ, McCall JM (1987) Novel 21-aminosteroids as potent inhibitors of iron-dependent lipid peroxidation. J Biol Chem 262: 10438–10440

Cadet JL, Katz M, Jackson-Lewis V, Fahn S (1989) Vitamin E attenuates the toxic effects of intrastriatal injection of 6-hydroxydopamine (6-OHDA) in rats: behavioral and biochemical evidence. Brain Res 476: 5–10

Cohen G, Werner P (1994) Free radicals oxidative stress and neurodegeneration. In: Calne DB (ed) Neurodegenerative disorders. Saunders, Philadelphia, pp 139–162

Colzi A, Turner K, Lee AJ (1998) Continuous subcutaneous waking day apomorphine in the long term treatment of levodopa induced interdose dyskinesias in Parkinson's disease. J Neurol Neurosurg Psychiatry 64: 573–576

Di Monte D, Dandy MS, Smith MT (1987) Increase efflux rather than oxidation is the mechanism of glutathione dspletion by 1-methyl-4-phenyl-1,2,3,6-tetrahydropyridine (MPTP). Biochem Biophys Res Commun 148: 153–160

Ebadi M, Srinivasan Sk, Baxi MD (1996) Oxidative stress and antioxidant therapy in Parkinson's disease. Prog Neurobiol 48: 1–19

Gancher S (1995) Pharmacokinetics of apomorphine in Parkinson's disease. J Neural Transm [Suppl] 45: 137–141

Gancher ST, Nutt JG, Woodward WR (1995) Apomorphine infusional therapy in Parkinson's disease: clinical utility and lack of tolerance. Mov Disord 10: 37–43

Gassen M, Youdim MBH (1998) Free radical scavengers: chemical concepts and clinical relevance. J Neural Transm (In press)

Gassen M, Glinka Y, Pinchasi B, Youdim MBH (1996) Apomorphine is highly potent free radical scavenger in rat mitochondrial fraction. Eur J Pharmacol 308: 219–225

Gassen M, Gross A, Youdim MBH (1998a) Apomorphine enantiomers protect pheochromocytoma (PC12) cells from oxidative stress induced by hydrogen peroxide and 6-hydroxydopamine. Mov Disord 13: 242–248

Gassen M, Pinchasi B, Youdim MBH (1998b) Apomorphine is a potent radical scavenger and protects culture pheochromocytoma cells from 6-OHDA and H_2O_2-induced cell death. Adv Pharmacol 42: 320–324

Gerlach M, Ben-Shachar D, Riederer P, Youdim MBH (1994) Altered brain metabolism of iron as a cause of neurodegenerative diseases? J Neurochem 63: 793–806

Glinka Y, Youdim MBH (1995) Inhibition of mitochondrial complex I and IV by 6-hydroxydopamine. Euro J Pharmacol 292: 329–332

Glinka Y, Tipton KF, Youdim MBH (1996) Nature of inhibition of mitochondrial respiratory complex I by 6-hydroxydopamine. J Neurochem 66: 2004–2010

Glinka Y, Gassen M, Youdim MBH (1997) Mechanism of 6-hydroxydopamine neurotoxicity. J Neural Transm [Suppl] 50: 55–66

Greene LA, Aletta JM, Rukenstein A, Green SH (1987) PC12 pheocytoma cells, culture, nerve growth factor treatment and experimental exploitation. Methods Enzymol 147: 207–216

Heikkila RE, Manzino L, Cabbat FS, Duvoisin RC (1984) Protection against the dopaminergic neurotoxicity of 1-methyl-4-phenyl-1,2,3,6-tetrahydropyridine by monoamine oxidase inhibitors. Nature 311: 467–469

Jellinger K, Paulus W, Grundke-Iqbal I, Riederer P, Youdim MBH (1990) Brain iron and ferritin in Parkinson's and Alzheimer's diseases. J Neural Transm [PD Sect] 2: 327–340

Jenner P, Olanow CW (1996) Pathological evidence for oxidative stress in Parkinson's disease and related degenerative disorders. In: Olanow CW, Jenner P, Youdim MBH (eds) Neurodegeneration and neuroprotection in Parkinson's disease. Academic Press, London, pp 24–45

Kato T, Horiuchi S, Togari A, Nagatsu T (1981) A sensitive and inexpensive high-performance liquid chromatographic assay for tyrosine hydroxylase. Experientia 37: 809–811

Lan J, Jiang DH (1997) Desferrioxamine and vitamin E protect aginst iron and MPTP-induced neurodegeneration in mice. J Neural Transm 104: 469–481

Lees AJ (1993) Dopamine agonists in Parkinson's disease: a look at apomorphine. Fundam Clin Pharmacol 7: 121–128

LeWitt PA (1994) Clinical trials of neuroprotection in Parkinson's disease: long-term selegiline and alpha-tocopherol treatment. J Neural Transm [Suppl] 43: 171–181

Linert W, Herlinger E, Jameson RF, Kienzl E, Jellinger K, Youdim MBH (1996) Dopamine, 6-hydroxydopamine, iron, and dioxygen-their mutual interactions and possible implication in the development of Parkinson's disease. Biochem Biophys Acta 1316: 160–168

Liu J, Mori A (1993) Monoamine metabolism provides an antioxidant defense in the brain against oxidant- and free radical-induced damage. Arch Biochem Biophys 302: 118–127

Matthews RT, Yang L, Beal M (1997) S-Methylthiocitrulline, a neuronal nitric oxide synthase inhibitor, protects against malonate and MPTP neurotoxicity. Exp Neurol 143: 282–286

Mochizuki H, Imai H, Endo K, Yokomizo K, Murata Y, Hattori N, Mizuno Y (1994) Iron accumulation in the substantia nigra of 1-methyl-4-phenyl-1,2,3,6-tetrahydropyridine (MPTP) induced hemiparkinsonism in monkeys. Neurosci Lett 168: 251–253

Monteiro HP, Winterbourn CC (1989) 6-Hydroxydopamine releases iron from ferritin and promotes ferritin-dependent lipid peroxidation. Biochem Pharmacol 38: 4177–4182

O'Carroll AM, Fowler CJ, Phillips JP, Tobbia I, Tipton KF (1983) The deamination of dopamine by human brain monoamine oxidase. Specificity for the two enzyme forms in seven brain regions. Naunyn Schmiedebergs Arch Pharmacol 322: 198–202

Oestreicher E, Sengstock GJ, Riederer P, Olanow CW, Dunn AJ, Arendash GW (1994) Degeneration of nigrostriatal dopaminergic neurons increases iron within the substantia nigra: a histochemical and neurochemical study. Brain Res 660: 8–18

Olanow CW (1993) A scientific rationale for protective therapy in Parkinson's disease. J Neural Transm [Gen. Sect] 91: 161–180

Olanow CW, Youdim MBH (1996) Iron and neurodegeneration: Prospects for neuroprotection. In: Olanow CW, Jenner P, Youdim MBH (eds) Neurodegeneration and neuroprotection in Parkinson's disease. Academic Press, New York, pp 50–67

Perumal AS, Gopal VB, Tordzro WK, Cooper TB, Cadet JL (1992) Vitamin E attenuates the toxic effects of 6-hydroxydopamine on free radical scavenging systems in rat brain. Brain Res Bull 29: 699–701

Przuntek H (1994) Clinical aspects of neuroprotection in Parkinson's disease. J Neural Transm [Suppl] 43: 163–169

Rogers JT, Bridges KR, Durmowicz GP, Glass J, Zuron PE, Munro HN (1990) Translational control during the acute phase response. Ferritin synthesis in response to interleukin-1. J Biol Chem 265: 14572–14578

Sam EE, Verbeke N (1995) Free radical scavenging properties of apomorphine enantiomers and dopamine — possible implication in their mechanism of action in parkinsonism. J Neural Transm [PD Sect] 10: 115–127

Sam E, Jeanjean AP, Maloteaux JM, Verbeke N (1995) Apomorphine pharmacokinetics in parkinsonism after intranasal and subcutaneous application. Eur J Drug Metab Parmacokinet 20: 27–33

Santiago M, Matarredona ER, Granero L, Cano J, Machado A (1997) Neuroprotective effect of the iron chelator desferrioxamine against MPP$^+$ toxicity on striatal dopaminergic terminals. J Neurochem 68: 732–738

Seiser C, Teixeira S, Kuhn LC (1993) Interleukin-2 dependent transcriptional and post-transcriptional regulation of transferrin receptor mRNA, J Biol Chem 268: 13074–13080

Sengstock GJ, Olanow CW, Dunn AJ, Barone S Jr, Arendash GW (1994) Progressive changes in striatal dopaminergic markers, nigral volume, and rotational behavior following iron infusion into the rat substantia nigra. Exp Neurol 130: 82–94

Simeonova PP, Luster MI (1995) Iron and reactive oxygen species in the asbestos-induced tumor necrosis factor-alpha response from alveolar macrophages. Am J Respir Cell Mol Biol 12: 676–683

Singer TP, Castagnoli N Jr, Ramsay RR, Treor AJ (1987) Biochemical events in the development of parkinsonism induced by 1-methyl-4-phenyl-1,2,3,6-tetrahydro-pyridine. J Neurochem 49: 1–8

Sorrenti V, Di Giacomo C, Renis M, Russo A, La Delfa C, Perez-Polo JR, Vanella A (1994) Lipid peroxidation and survival in rats following cerebral post-ischemic reperfusion: effect of drugs with different molecular mechanisms. Drugs Exp Clin Res 20: 185–189

Stadtman ER (1993) Oxidation of free amino acids and amino acid residues in proteins by radiolysis and by metal-catalyzed reactions. Annu Rev Biochem 62: 797–821

Temlett JA, landsberg JP, Watt F, Grime GW (1994) Increased iron in the substantia nigra compacta of the MPTP-lesioned hemiparkinsonian african green monkey: evidence from proton microprobe element microanalysis. J Neurochem 62: 134–146

Vimard F, Nouvelot A, Duval D (1996) Cytotoxic effects of oxidative stress on neuronal-like pheochromocytoma cells. Biochem Pharmacol 51: 1389–1395

Youdim MBH, Ben-Shachar D, Riederer P (1990) The role of monoamine oxidase, iron-melanin interaction, and intracellular calcium in Parkinson's disease. J Neural Transm [Suppl] 32: 239–248

Youdim MBH, Ben-Shachar D, Riederer P (1993) The possible role of iron in the etiopathology of Parkinson's disease. Mov Disord 8: 1–12

Zhao W, Richardson JS, Mombourquette MJ, Weil JA (1995) An in vitro EPR study of the free-radical scavenging action of the lazaroid antioxidants U-74500A and U-78517F. Free Radic Biol Med 19: 21–30

Zigmond MJ, Berger TW, Grace AA, Stricker EM (1989) Compensatory responses to nigrostriatal bundle injury. Mol Chem Neuropathol 10: 185–200

Authors' address: Prof. M. B. H. Youdim, B. Rappaport Family Medical Science Building, Efron Street, P.O.B. 9697, Haifa 31096, Israel

Neuroprotective effect of chronic inactivation of the subthalamic nucleus in a rat model of Parkinson's disease

B. Piallat, A. Benazzouz, and **A. L. Benabid**

Laboratoire de Neurobiologie Préclinique, INSERM U.318, CHU, Grenoble, France

Summary. Several evidences showed that glutamate can be implicated in the degenerative process of dopaminergic neurons in Parkinson's disease. The treatment with NMDA antagonists have been shown to induce a neuroprotective effect in animal models of this disease. As subthalamic nucleus neurons send direct glutamatergic projections to the substantia nigra, we studied the effects of kainic acid lesion of this nucleus on the degeneration of dopaminergic neurons induced by microinjection of 6-hydroxydopamine in the striatum of rat done one week after the first lesion. Animals were killed 15 days after the injection of 6-hydroxydopamine. Immunohistochemical study showed that lesion of the subthalamic nucleus can prevent the degeneration of substantia nigra dopaminergic somata when carried out one week prior to 6-hydroxydopamine injection in the striatum. Nevertheless neurochemical results showed that this lesion did not antagonize the striatal 6-hydroxydopamine-induced dopamine depletion in the striatum 15 days after 6-hydroxydopamine injection.

Introduction

The subthalamic nucleus (STN) has been demonstrated to play a key role in the control of movement. Its hypoactivity by in situ injection of GABA agonists or its lesion are related to the clinical appearence of involuntary dyskinetic movements (Whittier, 1947; Whittier and Mettler, 1949; Crossman, 1987). During the last decade, STN has been described to play a glutamatergic excitatory influence on the principal output nuclei of the basal ganglia system (Kitai and Kita, 1987; Robledo and Féger, 1990). In MPTP-treated monkey, it has been shown that STN is overactive (Miller and DeLong, 1987; Bergman et al., 1994) and that its lesion can reverse motor parkinsonian symptoms (Bergman et al., 1990; Aziz et al., 1991). Nevertheless, this beneficial effect was accompanied by the appearence of abnormal dyskinetic movements. More recently, we have shown that bradykinesia and rigidity can be alleviated by the application of electrical high frequency stimulation at STN level in hemiparkinsonian monkey (Benazzouz et al., 1993; 1996) and in akineto-rigid parkinsonian patients (Limousin et al., 1995a; 1995b). Concerning the functional mechanism of this type of stimulation, we have shown that high frequency stimulation of STN induced an inhibition of neuronal activity in the two main output structures of basal ganglia in rats: the pars reticulata of substantia nigra (SNr) and the entopeduncular nucleus (EP, the equivalent structure of internal pallidum in primate) (Benazzouz et al., 1995). The decrease of activity in these two nuclei is due to a direct inhibition of STN glutamatergic neurons probably by a depolarization block. Then experimental data provide evidence that STN neuroinhibition is achieved through a shutdown of the glutamatergic output of STN cells, while the basic phenomenon is still not demonstrated. This shutdown of the glutamate outflow may have potential effects on the development and time course of the degenerative process of dopaminergic cells.

Several studies suggested that glutamate excitotoxicity mediated by the activation of NMDA receptors could be implicated in the pathogenesis of Parkinson's disease and that NMDA antagonists can be used as neuroprotective agents in animal models of the disease. Authors found that both competitive and non competitive NMDA antagonists protected against the lesion of dopaminergic neurons induced by methamphetamine in rats (Sonsalla et al., 1991; O'Dell et al., 1992; Marshall et al., 1993). The enhanced neurotoxicity of methamphetamine, caused by unilateral administration of NMDA into the striatum, has confirmed the role of NMDA receptors in toxic mechanisms (Sonsalla et al., 1989). Turski et al. (1991) presented some evidence for the concept of participation of NMDA receptors in 1-methyl-4-phenylpyridinium (MPP^+) neurotoxicity in rats. They showed that the administration of non competitive NMDA antagonists (MK-801 or AP7) prevented the depletion of dopamine induced by the injection of MPP^+. Moreover, Storey et al. (1992) have shown that intrastriatal MPP^+ lesions were partially blocked by MK-801 systemic administration. Moreover, MK-801 administered to monkeys, jointly with MPTP prevented degeneration of dopaminergic neurons (Zuddas et al., 1992).

Why STN inactivation can play a role in the protection of substantia nigra dopaminergic neurons?

It is clear that STN plays an important role in the manifestation of parkinsonian symptoms and that its glutamatergic neurons exert an excitatory influence on both dopaminergic and non dopaminergic cells in the substantia nigra. (Hammond et al., 1978; Parent et al., 1995; Robledo and Féger, 1990). As SNc dopaminergic neurons contain NMDA receptors, we wanted to determine if the inactivation of STN could prevent or stop the degeneration of dopaminergic neurons exposed to a selective neurotoxin.

Material and methods

A group of rats (Group 1: KA-STN/6-OHDA, n = 20) received unilateral injection of kainic acid (KA, 2 µg dissolved in 0.5 µl saline) into the STN and one week later an injection of 6-OHDA (20μg dissolved in 4μl of saline) into the striatum. The control group (Group 2, 6-OHDA, n = 20) received only 6-OHDA into the striatum and the animals of group 3 (sham, NaCl-STN/6-OHDA, n = 20) was injected by saline (NaCl 0.9%) into the STN and one week later 6-OHDA into the striatum. Animals were observed and then killed 15 days after the injection of 6-OHDA. The animals of each group were separated into 2 subgroups (n = 10 each subgroup). Rats of the first subgroup were used for immunohistochemical study to determine the percentage of dopaminergic cell loss using immunohistochemistry of Tyrosine Hydroxylase (TH) as previously described (Piallat et al., 1996). For this, animals were deeply anaesthetized with chloral hydrate and transcardially perfused with 4% paraformaldehyde. The animals of the second subgroup were used for biochemical study and were killed by decapitation. Brains were quickly removed from the skull, rinsed in saline. Lesioned and unlesioned striata were separated, dissected and then frozen. The tissue samples were homogenized, centrifuged and the supernatant was filtered and stored at −80°C. Analysis of dopamine was performed using high performance liquid chromatography (HPLC).

Behavioral results

The results of the behavioral study showed that after awakening from anaesthesia, all rats who received kainic acid injection into the STN exhibited repeated involuntary movements of anterior and posterior legs. Then, they showed spontaneous contralateral rotational behaviour and locomotor hyperactivity. In these rats, before and after 6-OHDA injection, apomorphine treatment induced behavioral rotations ipsilateral to the lesioned side. In the contrary, rats who received only 6-OHDA injection into the striatum presented apomorphine-induced contralateral rotational behaviour.

Immunohistochemical results

The analysis of immunohistological brain sections, after counting dopaminergic cell bodies in the substantia nigra, showed (Fig. 1) that STN lesion induced a neuroprotective effect on dopaminergic cells when carried out one week prior to 6-OHDA injection (Piallat et al., 1996). Statistical analysis did not show significant differences in the number of TH-immunoreactive (TH-IR) cell bodies between normal side and lesioned side. The percentage of SNc dopaminergic neuron loss was very low (7.6 ± 3.3%). In contrast, in rats of the control group (who received only intrastriatal injection of 6-OHDA) and sham group who received saline in the STN prior to 6-OHDA in the striatum, the percentage of TH-IR cell loss in the SNc on lesioned side was 43.5 ± 5.2% and 40.7 ± 7.8% respectively (Fig. 1).

Biochemical results

The analysis of results showed no significant difference in dopamine level between intact sides of all groups of rats compared to the values obtained in normal rats (Fig. 2). Fifteen days after microinjection of 6-OHDA into the

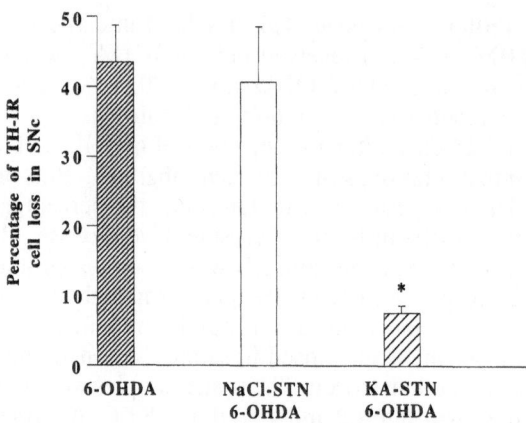

Fig. 1. Histograms showing the percentage of SNc cell loss in the lesioned sides of the different groups of rats compared to the contralateral sides. *6-OHDA* group of rats receiving 6-OHDA injection in the striatum; *NaCl-STN/6-OHDA* group of rats receiving saline (NaCl 0.9%) in the STN and 7 days later 6-OHDA into the striatum; *KA-STN/6-OHDA* group of rats receiving kainic acid in the STN and 7 days later 6-OHDA into the striatum. *indicates significant difference (Student's t-test, $p < 0.05$) in the percentage of cell loss between rats of group KA-STN/6-OHDA and those of two other groups. NS indicates non significant difference between rats of group NaCl-STN/6-OHDA and those of group 6-OHDA

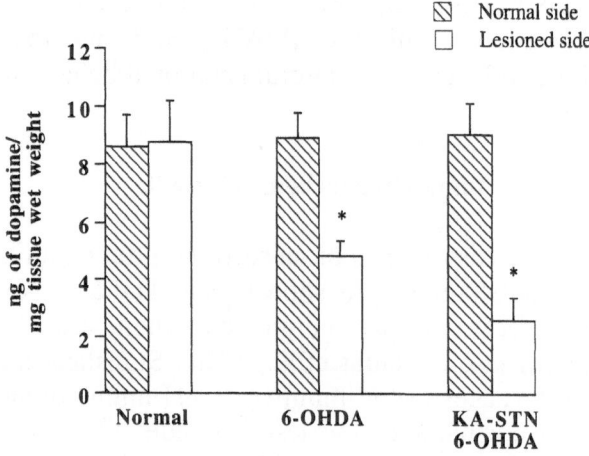

Fig. 2. Histograms showing the effects of 6-OHDA and combined treatment of 6-OHDA and STN kainic acid lesion on dopamine levels in the striatum of rats 15 days after the microinjection of 6-OHDA into the striatum. *indicates significant difference (Student's t-test, $p < 0.05$) in the dopamine level between normal side and lesioned side of each group of rats. NS indicates non significant difference in the dopamine level between the two sides of normal rats

striatum, a significant decrease of dopamine level was observed in the lesioned side compared to the normal side of rats with or without STN lesion (Fig. 2). No significant difference in dopamine level was observed in these two groups.

Discussion

The results of the present work show that STN lesion, when done one week before 6-OHDA injection, can prevent the degeneration of nigral dopaminergic cells in this rat model of Parkinson's disease. Nevertheless, this lesion did not modify the striatal dopamine depletion after microinjection of 6-OHDA into the striatum. This discrepancy of biochemical and immunohistochemical results is not very surprising because similar results were shown in other models of Parkinson's disease (MPTP-treated mice) prevently treated with NMDA antagonists (Löschmann et al., 1994) or with a selective blocker of calcium channels, Nimodipine (Kupsch et al., 1995). Our results can be explained by the fact that 6-OHDA could exert its toxic effect in nigro-striatal dopaminergic terminals reducing the level of dopamine in the striatum which is comparable to that obtained in rats receiving only 6-OHDA in the striatum. This result is very important to consider because it suggests that, in this model, 6-OHDA starts its toxic mechanism at striatal level and then this toxicity progress to the somata in the SNc. The quantification of striatal dopamine was done 15 days after 6-OHDA injection. When this quantification was done 90 days after 6-OHDA injection, dopamine level in the treated side was not significantly different from that obtained in normal side or in both sides of normal rats (data not shown). The immunohistological results showing that STN lesion can protect somata of SNc against 6-OHDA can be explained by the fact that kainic acid lesion provokes a shutdown of glutamate release in the SNc. Combined immunohistochemical and neurochemical methods show that tyrosine hydroxylase immunoreactive loss and DA level reduction have different time courses, cell terminals being first injured inducing retrograde degeneration, and then the somata are functionally inhibited and then they die if they are not rescued by the glutamate shutdown following STN inhibition.

In conclusion the present study shows that destruction of excitatory glutamatergic neurons of the STN protects the SNc dopaminergic neurons in a rat model of progressive parkinsonism. The neurochemical level of striatal dopamine is, in a first time, reduced due to the 6-OHDA-induced cell terminals toxicity and then a delayed functional improvement of dopamine level is obtained with a mechanism which remained unclear, probably by a regeneration of cell terminals of the surviving SNc dopaminergic neurons. These results open a new field in the therapeutic strategies of Parkinson's disease suggesting that surgical or pharmacological inactivation of the STN could stop or slow down the progression of this disease.

References

Aziz TZ, Peggs D, Sambrook MA, Crossman AR (1991) Lesion of the subthalamic nucleus for the alleviation of protection protection protection 1-methyl-4-phenyl-1,2,3,6-tetrahydropyridine (MPTP)-induced parkinsonism in the primate. Mov Disord 6: 288–292

Benazzouz A, Gross C, Féger J, Boraud T, Bioulac B (1993) Reversal of rigidity and improvement in motor performance by subthalamic high-frequency stimulation in MPTP-treated monkeys. Eur J Neurosci 5: 382–389

Benazzouz A, Piallat B, Pollak P, Benabid AL (1995) Responses of substantia nigra pars reticulata and globus pallidus complex to high frequency stimulation of the subthalamic nucleus in rats: electrophysiological data. Neurosci Lett 189: 77–80

Benazzouz A, Boraud Th, Féger J, Burbaud P, Bioulac B, Gross Ch (1996) Alleviation of experimental hemiparkinsonism by high frequency stimulation of the subthalamic nucleus in primate: A comparison with L-dopa treatment. Mov Disord 11: 627–632

Bergman H, Wichmann T, Delong MR (1990) Reversal of experimental parkinsonism by lesions of the subthalamic nucleus. Science 249: 1436–1438

Bergman H, Wichmann T, Karmon B, DeLong MR (1994) The primate subthalamic nucleus. II. Neuronal activity in the MPTP model of parkinsonism. J Neurophysiol 72: 507–520

Hammond C, Deniau JM, Rizk A, Féger J (1978) Electrophysiological demonstration of an excitatory subthalamonigral pathway in the rat. Brain Res 151: 235–244

Limousin P, Pollak P, Benazzouz A, Hoffmann D, Lebas JF, Broussole E, Perret JE, Benabid AL (1995a) Effect on parkinsonian signs and symptoms of bilateral subthalamic nucleus stimulation. The Lancet 345: 91–95

Limousin P, Pollak P, Benazzouz A, Hoffmann D, Broussolle JE, Perret J, Benabid AL (1995b) Bilateral subthalamic nucleus stimulation for severe Parkinson's disease. Mov Disord 10: 672–674

Marshall JF, O'Dell SJ, Weihmuller FB (1993) Dopamine-glutamate interactions in methamphetamine-induced neurotoxicity. J Neural Transm 91: 241–254

Miller WC, De Long MR (1987) Altered tonic activity of neurons in the globus pallidus and subthalamic nucleus in the primate MPTP model of parkinsonism. In: Carpenter MB, Jayaraman A (eds). The basal ganglia II, Advances in Behavioral Biology 32 Plenum, New York pp 415–427

Mitchell IJ, Clarke CE, Boyce S, Robertson RG, Peggs D, Sambrook MA, Crossman AR (1989) Neural mechanisms underlying parkinsonian symptoms base upon regional uptake of 2-deoxyglucose in monkeys exposed to 1-methyl-4-phenyl-1,2,3,6-tetrahydropyridine. Neurosci 32: 213–226

O'Dell SJ, Weihmuller FB, Marshall JF (1992) MK-801 prevents methamphetamine-induced striatal dopamine damage and reduces extracellular dopamine overflow in neurotoxins and neurodegenerative disease. Ann NY Acad Sci 648: 317–319

Parent A, Hazrati L (1995) Functional anatomy of the basal ganglia. II. The place of subthalamic nucleus and external pallidum in basal ganglia circuitry. Brain Res Rev 20: 128–154

Piallat B, Benazzouz A, Benabid AL (1996) Subthalamic nucleus lesion in rat prevents dopaminergic nigral neuron degeneration after striatal 6-OHDA injection: behavioral and immunohistochemical studies. Eur J Neurosci 8: 1408–1414

Robledo P, Féger J (1990) Excitatory influence of rat subthalamic nucleus to substantia nigra pars reticulata and the pallidal complex: electrophysiological data. Brain Res 518: 47–54

Sonsalla PK, Nicklas WJ, Heikkila RE (1989) Role for excitatory amino acids in methamphetamine induced nigrostriatal dopaminergic toxicity. Science 243: 398–400

Sonsalla PK, Riordan DE, Heikkila RE (1991) Competitive and non competitive antagonists at NMDA receptors protect against methamphetamine-induced dopaminergic damage in mice. J Pharmacol Exp Ther 256: 506–512

Storey E, Hyman BT, Jenkins B, Bouillet E, Miller JM, Rosen BR, Beal FM (1992) MPP+ produces excitotoxic lesions in rat striatum as a result of impairment of oxidative metabolism. J Neurochem 58: 1975–1978

Turski L, Bressler K, Rettig KJ, Löschmann PA, Watchel H (1991) Protection of substantia nigra from MPP+ neurotoxicity by NMDA antagonists. Nature 349: 414–418

Zuddas A, Oberto G, Vaglini F, Fascetti F, Fornai F, Corsini GU (1992) MK-801 prevents MPTP-induced parkinsonism in primates. J Neurochem 59: 133–739

Authors' address: A. L. Benabid, MD, Laboratoire de Neurobiologie Préclinique, INSERM U.318, CHU — Pavillon B, BP 217, F-38043 Grenoble Cedex 09, France

Antiglutamate therapy of ALS — which is the next step?

A. C. Ludolph, T. Meyer, and **M. W. Riepe**

Department of Neurology, University of Ulm, Federal Republic of Germany

Summary. Amyotrophic lateral sclerosis (ALS) is a fatal neurodegenerative disease which was thought to be untreatable for a long time. However, recent evidence in men indicates that antiglutamatergic strategies are the first to have an influence on its pathogenesis and slow down the disease process. Since the effect of the drugs is still small, this progress cannot only be seen as a success of the present but most also be acknowledged as a starting point for the future. How will these future studies look like? They will have to take into account that ALS presumably has a long preclinical period and they will use a number of novel compounds and treatment strategies which have recently been shown to be effective in a transgenic animal model. This also implies that we are likely to use combination therapies and have to try to treat patients early. The latter will be necessarily connected with the demand for a novel clinical attitude to the diagnosis of the disease.

Introduction

Since the first description by Charcot in the late nineteenth century (Charcot, 1874), amyotrophic lateral sclerosis (ALS) was seen as a mysterious, untreatable and hopeless condition. The progressive, almost exclusively motor disease has a focal onset, spreads to every region of the body, but spares oculomotor nuclei and motor sphincter function. ALS predominantly affects men, has peak incidence rates between age 50–70, and life expectancy after diagnosis is on average not longer than 1–2 years (World Federation of Neurology Research Group on Neuromuscular Diseases, 1994). In about 10% of the patients the disease is inherited in an autosomal-dominant fashion. Death results in most patients from respiratory failure. In the past, early diagnosis was not regarded to be an advantage for the patient and as soon as the disease was diagnosed, therapeutic nihilism dominated the clinical field. A large number of, partly poorly designed, therapeutical trials remained unsuccessful and hypotheses were the ground on which pathogenetic ideas were founded. Substantial progress was made after the discovery of the first gene responsible for the etiology of the disease in the subgroup of patients suffering from the autosomal-dominant familial form of amyotrophic lateral sclerosis (fALS) (Rosen et al., 1993). The description of mutations in the gene for the cytosolic form of the Cu/Zn superoxide dismutase (Cu/Zn SOD) fueled hopes that this specific etiology also explains the pathogenesis by a reduction of detoxification mechanisms for cytotoxic oxygen free radicals (Rosen et al., 1993). However, studies of transgenic animals carrying multiple copies of the human mutated gene clearly showed that not a loss of function but a toxic gain of function of the enzyme must explain the pathogenesis of the disease (Gurney et al., 1997; Wong et al., 1995). Early ultrastructural studies of the animals showed that mitochondrial damage is the hallmark of the early pathogenesis, later motor neurons disappear, and finally animals reach the stage of a clinically apparent disease with paraparesis and tetraparesis (Dal Canto and Gurney, 1994; Kong and Xu, 1998; Wong et al., 1995). Similar to human patients, mice finally die from respiratory failure. In a short time frame, the transgenic model has served to generate well-founded hypotheses on the pathogenesis of human familial and sporadic ALS (sALS), to improve our understanding of the sequence of events leading to selective vulnerability in the motor system, and — possibly most importantly — to screen and examine novel treatment strategies to rescue motor neurons from premature death.

The effects of antiexcitotoxic strategies in men and mice

The first treatment strategies successful in mice and men used drugs which reduced the toxic effects of excitatory amino acids. These strategies are based on two concepts of neuronal death, the concepts of direct and indirect excitotoxicity.

Direct and indirect excitotoxicity

The concepts of excitotoxicity are one possible explanation to understand selective neuronal death in the central nervous system of mammals. Since a number of drugs interfering with this mechanism of cell death are available or in development, this concept is of direct relevance for future therapeutic approaches:

Lucas and Newhouse in the fifties and Olney in the late sixties showed that increased extracellular levels of excitatory neurotransmitters such as glutamate can damage postsynaptic cells (direct excitotoxicity) (Lucas and Newhouse, 1957; Olney, 1969a; Olney, 1969b). This damage has distinct morphological features: it largely spares axons and glial cells, but affects the cell soma and the dendritic tree. It has been shown that this pattern of damage mirrors the distribution of cellular glutamate receptors indicating that glutamate induced damage is mediated by specific receptors. By the use of specific antagonists for these receptors it could be convincingly shown that this morphological pattern is indeed caused by receptor-mediated damage (Choi, 1988). Excitatory amino acids bind to postsynaptic receptors which leads to an opening of the associated ion channels, and sodium, water and calcium will enter the neuron. Increasing intracellular calcium levels will then activate a series of events which eventually leads to neuronal cell death. The mechanism of direct excitotoxicity has been shown to be relevant for a few exogenous diseases and its importance is widely acknowledged for acute diseases of the nervous system such as stroke, epilepsy, and brain trauma (Choi, 1988).

A more common part of the pathogenesis of cell death of largely energy-dependent neurons is the concept of indirect excitotoxicity (Henneberry et al., 1989). If a variety of etiologically relevant factors impair chemical energy synthesis in the neuron, comparatively low extracellular concentrations of excitatory amino acids are necessary to further disturb ion homoeostasis, depolarize the cell membrane potential to a critical level, and initiate cascades of biochemical events which lead to the morphologically defined phenomena of necrosis or apoptosis (Ankercrona et al., 1995; Beal, 1992; Henneberry et al., 1989; Leist and Nicotera, 1998; Ludolph et al., 1993). Studies of the membrane potential of the cell have shown that excitatory amino acids particularly drive depolarisation during initial stages of cellular energy depletion (Riepe et al., 1992; Riepe et al., 1994; Riepe et al., 1995). Consequently, antagonists to excitatory amino acid receptors have a larger protective effect on early phases of depolarisation compared with late ones (Riepe et al., 1995); in very late stages or after application of large dosages their effect is negligible.

Riluzole: mechanism of action

Riluzole (PK 26124, 2-amino-6-fluoromethoxy-benzothiazole) is a compound which interferes with excitatory neurotransmission and has anticonvulsive, sedative and neuroprotective properties in mammals (Doble, 1996; Doble, 1997; Malgouris et al., 1989; Mizoule et al., 1985). In particular, it has been shown to be efficient in a number of models for selective neuronal degeneration, including the MPP[+] model for Parkinson's disease (Benazzouz et al., 1995; Boireau et al., 1994; Bruijn et al., 1997) and the 3-nitropropionic acid model for Huntington's disease (Guyot et al., 1997). Although riluzole has a clear-cut effect on glutamatergic neurotransmisison, its mechanism of action is not completely understood and apparently has a number of facets (Doble, 1996; Doble, 1997). Riluzole does not directly interact with excitatory amino acid receptors at relevant concentrations but two molecular mechanisms seem to be of relevance:

1. Riluzole stabilizes voltage-dependent sodium channels in their inactivated state (Benoit and Escande, 1991; Stefani et al., 1997), possibly by directly binding to the alpha subunit of the channel (Hubert at al., 1994) and
2. The compound seems to activate a G-protein dependent signalling pathway since its effects are partly sensitive to pertussis toxin (Hubert et al., 1994).

These effects result in membrane stabilization and an apparently selective —
in comparison with other membrane-stabilizing compounds such as
lamotrigine — in vitro and in vivo effect on presynaptic glutamate release
(Cheramy et al., 1992; Martin et al., 1993). However, the data on selective
effects on release are not complete — it has been shown in vitro that GABA
and in vivo that aspartate release is not affected at comparable concentrations, but other transmitters or transmitter candidates have not been studied
yet.

In summary, it is certain that riluzole interferes with some specificity with
excitatory amino acid transmission in the central nervous system; the major
responsible molecular mechanisms are effects on voltage-dependent sodium
channels and G-protein dependent pathways. It may be that the low level of
side effects — if compared with other compounds interfering with excitatory
neurotransmission — is explained by the complexity of the mechanism of
action of riluzole.

Gabapentin: mechanism of action

Gabapentin is a structural analogue of the inhibitory neurotransmitter
gamma-amino butyric acid (GABA) which — in contrast to GABA itself —
easily passes the blood brain barrier. Although gabapentin is an established
anticonvulsant in both men and experimental animals, its mechanism of action remains far from being understood. The compound does not show affinity
to any of the major receptor groups — in particular the $GABA_A$ and $GABA_B$
receptors — but may have a specific, but yet unknown, binding site in rat
brain. Gabapentin also does not interact with sodium or potassium channels.
The drug inhibits the enzyme branched-chain amino acid transferase (BCAA-
T) causing a reduced biosynthesis of glutamate from alpha-ketoglutarate in
neurons and glial cells (Taylor, 1995). In addition, gabapentin seems to increase glutamate metabolism into gamma-amino butyric acid (Löscher et al.,
1991). By these mechanisms, gabapentin presumably reduces the intracellular
glutamate pool by both, inhibition of its biosynthesis and increased metabolism. This potentially reduces excitotoxic effects explaining its properties as an
anticonvulsant and potentially neuroprotective drug.

Riluzole and gabapentin in ALS

Both riluzole and gabapentin have been tried therapeutically in patients
suffering from amyotrophic lateral sclerosis. In 1994, in a stratified, double-
blind, placebo-controlled, prospective study Bensimon and collaborators
showed that after administration of 100 mg riluzole tracheostoma-free survival of ALS patients was increased significantly if compared with controls
(Bensimon et al., 1994). Although this was the first study to show an effect of
a neuroprotective drug in ALS, the effect on survival was restricted to — the
comparatively small number of — bulbar patients and conventional func-

tional scores did not show consistent protective effects. Therefore, a second large international study of 959 patients was initiated during which patients received in a dose-ranging design 25 mg, 50 mg, or 100 mg riluzole twice daily, or placebo (Lacomblez et al., 1996) for 18 months. Primary outcome measures were the same as in the first study. Again, Kaplan-Meier survival curves demonstrated prolonged survival in the group taking 100 mg riluzole; the same was shown for a dosage of 200 mg per day. Analysis of the four treatment groups revealed a dose response relationship (p = 0.04). If the Cox model of statistical analysis was used, each of the three treatment arms showed significant therapeutic effects. No significant effects were obtained when the secondary end points, such as manual muscle testing, Norris scores, clinical global impression and visual analogue scales were analyzed. It remains unclear why these measures did not show any efficacy; it may be that these non-linear and variable methods of assessment are too imprecise to detect differences of the magnitude seen here. Therefore, one of the lessons of the riluzole studies is that functional measures should be improved for use in future trials. Adverse effects were minor and the drug was well tolerated by the majority of patients.

In a placebo-controlled, double-blind phase II trial, the application of 2,400 mg gabapentin has been studied in 152 ALS patients (Miller et al., 1996) for six months. Standardized average maximum voluntary isometric strength from eight arm muscles served as the primary outcome measure whereas forced vital capacity was the secondary outcome measure. Statistical analysis showed that the decline of the primary outcome measure was reduced in a non-statistically significant manner (p = 0.057). Forced vital capacity was not significantly different between both groups (Miller et al., 1996). Because of this promising effect and since the drug was well tolerated, a larger second study was initiated which will be presumably completed in 1999.

Riluzole and gabapentin in transgenic mice

Both, riluzole and gabapentin, have effects on the survival of transgenic mice carrying multiple copies of the G93A mutation of the human SOD gene. Gurney and collaborators showed in 1996 that oral administration of riluzole (100 microgram/ml dissolved in water) increased life expectancy of this strain of mice by about 2 weeks (or 11%) (Gurney et al., 1996). Treatment was started in the preclinical period, at day 50 when the animals are still completely healthy (Gurney et al., 1996). Later, it could be shown that the effect was dose-dependent, and could also be detected by the use of functional scores. Similar, but smaller, effects were obtained after oral administration of gabapentin formulated at 3% in pelleted chow: survival was prolonged by about 8 days (Gurney et al., 1996). Gabapentin was also tried in a more benign form of the disease in mice which carried fewer transgene copies. This study confirmed the first studies and showed an increase in life expectancy of about 9 days (Gurney et al., 1996). Taken together, these studies demonstrate the therapeutic efficacy of these anticonvulsant and neuroprotective drugs in transgenic Cu/Zn SOD mice; indicating the usefulness of the therapeutic approach and suggesting that the mechanism of excitotoxicity is indeed part of the pathogenesis of selective anterior horn cell degeneration in these animals.

Are excitotoxic mechanisms part of the pathogenesis of motor neuron diseases?

There is only circumstantial evidence that direct excitotoxicity is part of the pathogenesis of human ALS; in comparison, arguments are stronger that indirect excitotoxicity is part of the pathogenesis of the currently best animal model, the Cu/Zn SOD transgenic animals. However, the question of the specificity of the latter findings is unresolved.

Human fALS and sALS

In postmortem studies of human ALS, glutamate levels were consistently found to be decreased in various regions of the brain, most likely reflecting a decreased intracellular glutamate pool (Malessa et al., 1991; Perry et al., 1987; Plaitakis et al., 1988; Tsai et al., 1990). Changes were not limited to those areas affected by the clinically apparent disease process. Presumably due to methodological limitations, results of numerous studies of extracellular glutamate concentrations were less consistent when measured in the patients' cerebrospinal fluid (CSF); increased concentrations may be found in a subset of patients (Shaw et al., 1995). In vitro studies showed that the CSF of ALS patients seems to contain a factor which is toxic in hippocampal cultures (Couratier et al., 1993). The neurotoxicity of this factor can be reduced by application of the non-NMDA antagonist CNQX, by riluzole, by vitamine E, and inhibitors of xanthine oxidase (Couratier et al., 1993; Couratier et al. 1994; Terro et al., 1996); results which seem to indicate that free radicals and excitotoxic mechanisms contribute to neuronal damage in this system. Studies of intra-, and extracellular glutamate homoeostasis seemed to imply the presence of an imbalance indicating that failure of high-affinity glutamate uptake systems could contribute to these finding. Consistent with this assumption, Rothstein and colleagues showed in synaptosomal preparations that the Na^+-dependent high-affinity glutamate uptake is deficient in affected regions of the central nervous system of ALS patients. The changes could not be detected in unaffected regions of ALS brain or in controls with Alzheimer's disease or Huntington's chorea (Rothstein et al., 1992). Later, this defect was shown to be due to low expression levels of the glial glutamate transporter subtype EAAT2 (GLT1) (Rothstein et al., 1995); however, the finding is still not completely consistent with the earlier biochemical studies which showed a reduction of intracellular glutamate levels in the entire brain. Mutation analysis of the complementary DNA for EAAT2 and EAAT3 isolated from motor cortex of a small number of ALS patients did not reveal disease-specific sequence alterations responsible for low expression levels (Meyer et al., 1995; Meyer et al., 1996); the same conclusion was drawn after larger studies which were based on the genomic sequence information (Aoki et al., 1998; Meyer et al., 1997; Meyer et al., 1998a; Meyer et al., 1998b). More recently, aberrant splicing of the EAAT2 transcript was suggested to be the cause for a reduced expression of the EAAT2 protein in ALS (Lin et al., 1998). However, it was shown by two groups that the alternative EAAT2 splice forms are expressed in ALS patients and controls alike suggesting that the aberrant splicing of these specific transcripts is an event which is not specific for ALS (Meyer et al., in press, Münch et al., 1998; Nagai et al., 1998). Therefore, at the present time it must be concluded that splicing of the EAAT2 transcript is unlikely to be ALS-specific and the regulation of EAAT2 protein expression is far being understood. Also, the large number of clinical studies of excitotoxicity in human ALS are far from presenting a conclusive and convincing picture although it seems to be of interest to pursue these studies.

Transgenic Cu/Zn SOD mice

There are also findings which may show that reduced expression of the EAAT2 protein contributes to the pathogenesis of anterior horn cell death in Cu/Zn SOD mice (Bruijn et al., 1997). In animals expressing a G85R-mutation of the human Cu/Zn superoxide dismutase, loss of EAAT2 expression has been detected (Bruijn et al., 1997). However, the same finding has been described in an ischemia model (Torp et al., 1995) and Alzheimer's disease (Masliah et al., 1996; Masliah et al., 1998) implying that aberrant functioning of EAAT2 may not be a specific part of the pathogenesis of this disorder.

Another finding strongly indicates that indirect excitotoxicity is part of the pathogenesis of this disorder in transgenic mice: Months after birth, these animals develop anterior horn cell depletion resulting — similar to the human disease — in a para- and tetraparesis (Bruijn et al., 1997; Chiu et al., 1995; Dal Canto and Gurney, 1994; Gurney et al., 1994; Wong et al., 1995). Much earlier, about 6–7 weeks after birth, high-copy strain animals develop early ultrastructural changes which consist of mitochondrial damage associated with vacuolation of the cell soma, dendrites and — in low copy animals — of the neuropil (Chiu et al., 1995; Kong et al., 1998; Wong et al., 1995). These changes resemble those seen in excitotoxic lesions (Ikonomidou et al., 1996). Although the specificity of these lesions is currently unknown (Burright et al., 1995), they remain important as long they are of therapeutic impact which seems to be the case (Gurney et al., 1996) as discussed above. Therefore, our currently best animal model for motor neuron degeneration supports the concept that treatment strategies which interrupt those cascades which follow mitochondrial damage, may they lead into apoptosis, may they lead into necrosis, will be of value to protect anterior horn cells.

The future: early treatment?

There is some evidence from mice and men that early treatment may lead to an improvement of the therapeutic effects currently seen.

A long preclinical period in Cu/Zn SOD mice

One of the major findings in the Cu/Zn SOD animals is the convincing evidence for the presence of a long preclinical period. Early ultrastructural changes of mitochondria and vacuolation of dendrites, axons and cell soma precede loss of anterior horn cells and the clinical symptomatology for a long time (Azzouz et al., 1997; Chiu et al., 1995; Kong and Xu, 1998; Wong et al., 1995) indicating that the cascade of events following these changes can be potentially interrupted by appropriate and early treatment (Kong and Xu, 1998). Motor neuron death seems to be restricted to the terminal stages of the disease (Kong and Xu, 1998) whereas early mitochondrial lesions can be demonstrated in adolescence or early adulthood of these animals (Chiu et al., 1995; Dal Canto and Gurney, 1994; Kong and Xu, 1998; Wong et al., 1995).

The effect of riluzole on early health states in ALS

In a retrospective analysis of the phase III riluzole study, it could be shown that the neuroprotective effect of the drug seems to be present in early ("mild health states"), but not in late stages of the disease (Riviere et al., 1998). Although this finding seems to indicate that early treatment of ALS should be the goal, the study has several drawbacks which reduce its significance. In particular, because of the inclusion criteria the size of the groups was very unevenly distributed favoring the presence of significant results in early in contrast to late health states (Ludolph and Riepe, in press).

Problems of early treatment

The idea of early diagnosis and treatment has been extensively discussed in acute neurological diseases such as stroke. In contrast, early diagnosis and treatment of neurodegenerative diseases is a new way of thinking which challenges the way neurologists handle the diseases and their patients. In other words, physicians must change their attitudes towards ALS patients if they accept that we need to diagnose and treat these diseases early. Which are the potential dangers if patients are treated early, but attitudes are not changed (Ludolph and Riepe, in press)?

Interaction with the patient

Pharmacological treatment must be complementary to other kinds of care such as supportive measures. It has been frequently stated that the opportunity of, in particular early, pharmacological neuroprotective treatment may lead to a reduction of psychological care. Only the combination of patient care — which includes any kind of symptomatic treatment and supportive measures — and neuroprotective drug treatment is the treatment of choice. It is unexpected news for a number of physicians that a drug, a neuroprotective drug, does not cure, but slows down a disease process. However, from the theoretical point of view such an effect is to be expected if a drug influences the pathogenesis, not the etiology of the disease in question. This experience is often made in medicine, for example in diseases such as arterial hypertension, certain kinds of cancer or diabetes. If physicians do not explain this basic principle of current — and possibly more efficient future — neuroprotective strategies to their patients, false hopes may be raised and consequently compliance may be small. Therefore, the attitude against treatment strategies which have the goal to reduce the speed of aging processes of specific neuronal cell populations must be changed and it needs to be accepted that current knowledge on the pathogenesis of ALS tells us that most steps and advances will be small. We accept the same in the therapy of arterial hypertension or other diseases of high prevalence.

If we promote early diagnosis and treatment we will also have to accept an increased rate of misdiagnoses. It will be impossible in some patients which present with atrophy of intrinsic hand muscles to decide with absolute certainty at this stage whether the patient develops rare Hirayama's disease or ALS. In the absence of a diagnostic gold standard for the early disease these mistakes are unavoidable and should be systematically controlled for. Also, an anticipated side effect of early diagnosis could be psychological reactions, such as depression or even suizide. These possible reactions must be included in our treatment strategies for the patient and seen as a potential and avoidable side effect. Also, if patients are treated early, they will be treated over an extended period of time. This increases the number of potential side effects which are to be controlled.

In summary, although a lot of arguments exist that future neuroprotective treatment of neurodegenerative diseases should be initiated early, this new way of thinking will also have some potential side effects which must and can be avoided.

Biological markers

If early or even preclinical diagnosis is to be made in the future to obtain an optimal therapeutic effect, biological markers which ascertain the diagnosis early would be of major help. Ideally, these markers should have some value to define the disease longitudinally. However, for ALS currently such markers are not available — with the exception of the Cu/Zn SOD mutations which obviously have a high specificity for the diagnosis. Proton spectroscopy presently seems to be useful to define motor cortex involvement in ALS by showing a reduction of NAA (N-acetylaspartate)/choline or NAA/creatine ratios, but does not seem to be able to define early stages of the disease (Block et al., 1998). Initial results (N = 9) seem to indicate that repeated analysis of the NAA /choline ratio could be a future quantifiable longitudinal marker for disease progression in the individual with regard to the upper motor neuron (Block et al., 1998). Another candidate biological marker might be deficient mitochondrial oxidative phosphorylation in muscle tissue of ALS patients (Wiedemann et al., 1998), if it could be observed non-invasively.

The future: novel treatment strategies

Based on preclinical studies in Cu/Zn SOD mice, a growing number of compounds and treatment strategies seem to be efficient in the attenuation of selective death of anterior horn cells.

Antioxidants

Vitamine E (tocopherol) is a major protective factor in lipid oxidation induced by free radicals. In models of cerebral infarction or brain trauma, rapid vitamine E depletion has been described. A dosage of 275 IU (200 mg supplement) vitamine E and 8.15 mg (8 mg supplement) selenium per kg diet has been used to examine the effect of these antioxidants on the course of the disease of the transgenic animals (Gurney et al., 1996). After given early during the life of these animals (day 50) this dosage delayed onset of clinical symptoms by about 12–15 days, but did not increase their life expectancy. This result was complemented by biochemical studies demonstrating that vitamine levels were 18fold increased in the liver of supplemented animals if compared to controls. This increase declined presumably due to reduced food intake during late stages of the disease. Also, an effect of vitamine E (in combination with selenium) on disease progression as demonstrated by running wheel activity could be shown. Carboxyfullerenes are malonic acid derivatives of buckminsterfullerenes (C_{60}) with potent antioxidant effects. These compounds have been administered to G93A transgenic mice at 73 ± 2 days of age by intraperitoneal mini-osmotic pumps (Dugan et al., 1997). The animals showed a delay of deterioration by about 10 days and died about 9 ($\pm$ 3.3) days later than controls indicating that carboxyfullerenes are also candidate neuroprotective drugs in motor neuron diseases.

In conclusion, it has been shown in two independent studies that treatment with antioxidants (both, vitamine E + selenium and carboxyfullerenes) has a neuroprotective effect in G93A Cu/Zn SOD transgenic mice indicating that such strategies might also be of benefit in the human disease. These results are complemented by recent findings in flies (*drosophila melanogaster*) which show that overexpression of the human Cu/Zn SOD in motor neurons of flies leads to an increase of the life span of about 40% if compared with controls (Parkes et al., 1998). This shows the importance of the enzyme for the function of motor neurons and the importance of motor neurons for the lifespan of this organism. Although it could be shown that Cu/Zn SOD overexpression made the fly more resistant to free radical challenge (Parkes et al., 1998), the Cu/Zn SOD is widely expressed in the nervous system and has a number of functions; therefore, it may not only be the detoxification of free radicals which contributes to this most interesting observation.

Antiapoptotic strategies

Cellular death by apoptosis (programmed cell death) is characterized by chromatin condensation, DNA fragmentation and the formation of apoptotic bodies. When apoptotic agents trigger cell death, cytochrome c is released from mitochondria into the cytoplasm and this activates a series of death-effector caspases, including interleukin-converting enzyme (ICE, caspase 1). It has been recently shown that Ca^{2+} accumulation in the mitochondrial matrix, reduced energy production, and a decrease of the mitochondrial membrane potential can trigger events which lead to apoptosis (Ankercrona et al., 1995; Leist and Nicotera, 1998). Therefore, it was consequent to employ neuroprotective strategies in models for motor neuron diseases which slow down or interrupt this sequence of events. Bcl-2 is a protein which can prevent apoptotic cell death by inhibiting the release of cytochrome c from mitochondria and — more downstream — also caspase activation directly (Hengartner, 1998). It has been shown if transgenic mice overexpressing Bcl-2 are cross-bred with transgenic mice carrying the mutant human Cu/Zn SOD, that the resulting strain has an increased life expectancy of about 30% (Kostic et al., 1997). The same has been shown by crossbreeding mutant Cu/Zn SOD mice with mice with a mutant and dysfunctional ICE (Friedländer et al., 1997).

In conclusion, these results show that not only the cascade of events which follows mitochondrial damage and leads to necrosis, but also the chain of events which leads to apoptosis — presumably in cells in which the membrane potential is more retained — can be interrrupted in order to obtain neuroprotective effects.

Others

D-penicillamine (D-β,β-dimethylcysteine) is a chelating drug which is used therapeutically in metal intoxications such as copper, mercury, zinc, and lead, but also in Wilson's disease. The compound also plays a therapeutical role in cystinuria and rheumatoid arthritis. Because of speculations that mutations in the Cu/Zn SOD gene may lead to an alteration of copper binding and secondarily increased, possibly neurotoxic, copper levels, d-penicillamine has been tried therapeutically in Cu/Zn SOD transgenic mice (Hottinger et al., 1997).

Administration of the drug early during the life of the animals led to an increase of life expectancy of more than 20% (Hottinger et al., 1997).

Consequences

Combination therapies

Since — like discussed above — a number of compounds and therapeutic principles have been successfully used in the currently best animal model for amyotrophic lateral sclerosis, combination therapy is one of the possible options of the future. It may be that the individual compound or therapeutic principle may have different effects on disease prevention, early or late stages of the disease. Therefore, combination therapy must possibly be refined by strategies which try to use the best drug combination at a specific time point of pathogenesis. Some combinations may also be more prone to side effects than others; for example combinations of antiexcitotoxic (membrane-stabilizing) drugs must presumably be monitored more carefully than combinations of drugs governed by different principles of action.

New forms of delivery

New forms of drug delivery are one of the challenges of the future for experimental medicine (Jain, 1998), in particular if large molecules are to be administered which have to cross the blood-brain barrier to reach the target tissue. Recently, a number of attempts have been made to overcome difficulties in bioavailability and to reduce toxicity. An obvious tool was the administration of growth factors such as brain-derived neurotrophic factor (BDNF) and ciliary derived neurotrophic factor (CNTF); compounds which initially were thought to be promising, but unfortunately were unsuccessful in large therapeutical trials for ALS, possibly because of a lack of bioavailability. Currently, trials on intrathecal application of BDNF and CNTF are underway (Ochs et al., 1998; Penn et al., 1997) and results can be expected in the near future. It seems that the pattern of side effects seen during intrathecal adminstration greatly differs from the ones seen after systemic application of these compounds (Ochs et al., 1998; Penn et al., 1997). Recently, adenoviral vectors have been successfully used to inject neurotrophin-3 (NT-3) intramuscularly into the mouse mutant pmn (progressive motor neuropathy). This therapeutic strategy increases the life span of the mutant by 50% (40.0 to 61.3 days), a result which can be further improved by coadministration of CNTF and NT-3 (66 days) (Haase et al., 1997). Similar results have been reported in a model of axotomized facial motor neurons (Gravel et al., 1997). However, for men these approaches for drug administration remain futuristic since a number of problems must be overcome until such a strategy can be applied in practice in medicine. The most important challenges to overcome are formation of antibodies to viruses, the development of antibodies to the respective growth factors, neurotoxicity, and the identification of specific target tissues (Sendtner, 1997).

Summary and conclusions

Since the riluzole studies were done, amyotrophic lateral sclerosis is seen as a treatable disease (Louvel et al., 1997). However, today the neuroprotective effects are still small and need to be improved. Studies of transgenic animals, and in a more limited way, the human disease, suggest that early diagnosis and treatment may improve the neuroprotective effect. The development of a biological disease marker which characterizes the disease longitudinally, ideally also in the preclinical period, would be a major step forward. It must not be disregarded that early diagnosis and treatment have potential side effects. Transgenic animals offer the opportunity to study a number of potential therapeutic neuroprotective approaches; the number of experimental drugs is rapidly increasing which promises that in a realistic time frame the riluzole effect can be improved by combination therapies. The choice of optimum drug combinations might differ for the different stages of the disease, for example early preclinical, early clinical and late clinical parts of the pathogenesis. In summary, although the first steps to neuroprotection have been done in ALS, they are still small, but developments in experimental therapy and new ideas on the pathogenesis of the disease offer some hope for better future treatment strategies of the disease. It must also not be forgotten that, if seen in perspective of the past, how encouraging current advances in the therapy of amyotrophic lateral sclerosis are.

References

Ankercrona M, Dypbukt JM, Bonfoco E, Zhivotovsky B, Orrenius S, Lipton SA, Nicotera P (1995) Apoptosis and mitochondria. Neuron 15: 961–973

Aoki M, Lin CL, Rothstein JD, Geller BA, Hosler BA, Munsat TL, Horvitz HR, Brown RH Jr (1998) Mutations in the glutamate transporter EAAT2 gene do not cause abnormal EAAT2 transcripts in amyotrophic lateral sclerosis. Ann Neurol 43: 645–653

Azzouz M, Leclerc N, Gurney M, Warter JM, Poindron P, Borg J (1997) Progressive motor neuron impairment in an animal model of familial amyotrophic lateral sclerosis. Muscle Nerve 20: 45–51

Beal MF (1992) Does impairment of energy metabolism result in excitotoxic neuronal death in neurodegenerative illnesses? Ann Neurol 31: 119–130

Benazzouz A, Boraud T, Dubédat P, Boireau A, Stutzmann JM, Gross C (1995) Riluzole prevents MPTP-induced parkinsonism in the rhesus monkey: a pilot study. Eur J Pharmacol 284: 299–307

Benoit E, Escande D (1991) Riluzole specifically blocks inactivated Na$^+$ channels in myelinated nerve fibers. Pflügers Arch 419: 603–609

Bensimon G, Lacomblez L, Meininger V and the ALS/Riluzole Study Group (1994) A controlled trial of riluzole in amyotrophic lateral sclerosis. N Engl J Med 330: 585–591

Block W, Karitzky J, Traber F, Pohl C, Keller E, Mundegar RR, Lamerichs R, Rink H, Ries F, Schild HH, Jerusalem F (1998) Proton magnetic resonance spectroscopy of

92   A. C. Ludolph et al.

the primary motor cortex in patients with motor neuron disease. Sub-group analysis and follow-up measurements. Arch Neurol 55: 931–936

Boireau A, Dubédat P, Bordier F, Peny C, Miquet JM, Durand G, Meunier M, Doble A (1994) Riluzole and experimental parkinsonism: antagonism of MPTP-induced decrease in central dopamine levels in mice. Neuroreport 5: 2657–2660

Bruijn LI, Becher MW, Lee MK, Anderson KL, Jenkins NA, Copeland NG, Sisodia SS, Rothstein JD, Borchelt DR, Price DL, Cleveland DW (1997) ALS-linked SOD1 mutant G85R mediates damage to astrocytes and promotes rapidly progressive disease with SOD1-containing inclusions. Neuron 18: 327–338

Burright EN, Clark HB, Servadlo A, Matilla T, Feddersen RM, Yunis WS, Duvick LA, Zoghbi HY, Orr HAT (1995) SCA1 transgenic mice: a model for neurodegeneration caused by an expanded CAG trinucleotide repeat. Cell 82: 937–948

Charcot JM (1974) Lecons sur les maladies du systéme nerveux faites à la Salpetrière. Paris. Progrès Medical 213–242

Cheramy A, Barbeito L, Godeheu G, Glowinsky J (1992) Riluzole inhibits the release of glutamate in the caudate nucleus of the cat in vivo. Neurosci Lett 147: 209–212

Chiu AY, Zhai P, Dal Canto MC, Peters TM, Kwon YW, Prattis TM, Gurney ME (1995) Age-dependent penetrance of disease in a transgenic mouse model of familial amyotrophic lateral sclerosis. Mol Cell Neurosci 6: 349–362

Choi DW (1998) Glutamate neurotoxicity and diseases of the nervous system. Neuron 1: 623–634

Couratier P, Hugon J, Sindou P, Vallat JM, Dumas M (1993) Cell culture evidence for neuronal degeneration in amyotrophic lateral sclerosis being linked to glutamate AMPA/kainate receptors. Lancet 341: 265–268

Couratier P, Sindou P, Esclaire F, Louvel E, Hugon J (1994) Neuroprotective effects of riluzole in ALS CSF toxicity. Neuroreport 5: 1012–1014

Dal Canto MC, Gurney ME (1994) Development of central nervous system pathology in a murine transgenic model of human amyotrophic lateral sclerosis. Am J Pathol 145: 1271–1279

Doble A (1996) The pharmacology and mechanism of action of riluzole. Neurology 47 [Suppl 4]: 233–241

Doble A (1997) Effects of riluzole on glutamatergic neurotransmission in the mammalian central nervous system, and other pharmacological effects. Rev Contemp Pharmacother 8: 213–225

Dugan LL, Turetsky DM, Du C, Lobner D, Wheeler M, Almli CR, Shen CK, Luh TY, Choi DW (1997) Carboxyfullerenes as neuroprotective agents. Proc Natl Acad Sci USA 94: 9434–9439

Friedländer RM, Brown RH, Gagliardini V, Wang J, Yuan J (1997) Inhibition of ICE slows ALS in mice. Science 388: 31

Gravel C, Götz R, Lorrain A, Sendtner M (1997) Adenoviral gene transfer of ciliary neurotrophic factor and brain-derived neurotrophic factor leads to long-term survival of axotomized motor neurons. Nat Med 3: 765–770

Gurney M, Pu H, Chiu A, Dal Canto M, Polchow C, Alexander D, Caliendo J, Hantati A, Kwon Y, Deng H, Chen W, Zhai P, Sifit R, Siddique T (1994) Motor neuron degeneration in mice that express a human Cu, Zn superoxide dismutase mutation. Science 264: 1773–1775

Gurney M, Cutting FB, Zhai P, Doble A, Taylor C, Andrus PK, Hall ED (1996) Benefit of vitamin E, riluzole, and gabapentin in a transgenic model of familial amyotrophic lateral sclerosis. Ann Neurol 39: 147–157

Guyot MC, Palfi S, Stutzmann JM, Maziere M, Hantraye P, Brouillet E (1997) Riluzole protects from motor deficits and striatal degeneration produced by systemic 3-nitropropionic acid intoxication in rats. Neuroscience 81: 141–149

Haase G, Kennel P, Pettmann B, Vigne E, Akli S, Revah F, Schmalbruch H, Kahn A (1997) Gene therapy of murine motor neuron disease using adenoviral vectors for neurotrophic factors. Nat Med 3: 429–436

Hengartner MO (1998) Death cycle and Swiss army knives. Nature 391: 441–442

Henneberry RC, Novelli A, Cox JA, Lysko PG (1989) Neurotoxicity at the N-methyl-D-aspartate receptor in energy-compromised neurons: an hypothesis for cell death in aging and disease. Ann NY Acad Sci 568: 225–233

Hottinger AF, Fine EG, Gurney ME, Zurn AD, Aebischer P (1997). The copper chelator d-penicillamine delays onset of disease and extends survival in a transgenic mouse model of familial amyotrophic lateral sclerosis. Eur J Neurosci 9: 1548–1551

Hubert JP, Delumeau JC, Prémont J, Glowinski J, Doble A (1994) Antagonism by riluzole of entry of calcium evoked by NMDA and veratridine in rat cultured granule cells: evidence for a dual mecahnism of action. Br J Pharmacol 113: 261–267

Ikonomidou C, Qin YQ, Labruyere J, Olney JW (1996) Motor neuron degeneration induced by excitotoxin agonists has features in common with those seen in the SOD-1 transgenic mouse model of amyotrophic lateral sclerosis. J Neuropathol Exp Neurol 55: 211–224

Jain RK (1998) The next frontier of molecular medicine: Delivery of therapeutics. Nature Med 4: 655–657

Kong J, Xu Z (1998) Massive mitochondrial degeneration in motor neurons triggers the onset of amyotrophic lateral sclerosis in mice expressing a mutant SOD1. J Neurosci 18: 3241–3250

Kostic V, Jackson-Lewis V, de Bilbao F, Dubois-Dauphin M, Przedborski S (1997) Bcl-2: prolonging life in a transgenic mouse model of familial amyotrophic lateral sclerosis. Science 277: 559–562

Lacomblez L, Bensimon G, Leigh PN, Guillet P, Meininger V (1996) Dose-ranging study of riluzole in amyotrophic lateral sclerosis. Lancet 347: 1425–1431

Leist M, Nicotera P (1998) Apotosis, excitoxicity, and neuropathology. Exp Cell Res 239: 183–201

Lin CLG, Bristol LA, Jin L, Dykes-Hoberg M, Crawford T, Clawson L, Rothstein JD (1998) Aberrant RNA processing in a neurodegenerative disease: the cause for absent EAAT2, a glutamate transporter, in amyotrophic lateral sclerosis. Neuron 20: 589–602

Löscher W, Hönack D, Taylor CP (1991) Gabapentin increases aminooxyacetic acid-induced GABA accumulation in regions of rat brain. Neurosci Lett 128: 150–154

Louvel E, Hugon J, Doble A (1997) Therapeutic advances in amyotrophic lateral sclerosis. TIPS 18: 196–203

Lucas DR, Newhouse JP (1957) The toxic effect of sodium L-glutamate on the inner layers of the retina. Arch Ophtalmol 58: 193–204

Ludolph AC, Riepe MW (1999) Do the benefits of currently available treatments justify early diagnosis and treatment of amyotrophic laterals sclerosis? Against, in press

Ludolph AC, Riepe M, Ullrich K (1993) Excitotoxicity, energy metabolism and neuronal degeneration. J Inher Metab Dis 16: 716–723

Malessa S, Leigh PN, Bertel O, Sluga E, Hornykiewicz O (1991) Amyotrophic lateral sclerosis: glutamate dehydrogenase and transmitter amino acids in the spinal cord. J Neurol Neurosurg Psychiatry 54: 984–988

Malgouris C, Bardot F, Daniel M, Pellis F, Rataud J, Uzan A, Blanchard C, Laduron RM (1989) Riluzole, a novel antiglutamate, prevents memory loss and hippocampal neuronal damage in ischemic gerbils. J Neurosci 9: 3720–727

Martin D, Thompson MA, Nadler JV (1993) The neuroprotective agent riluzole inhibits release of glutamate and asparatate from slices of hippocampal area CA1. Eur J Pharmacol 250: 473–476

Masliah E, Alford M, DeTeresa R, Mallory M, Hansen L (1996) Deficient glutamate transport is associated with neurodegeneration in Alzheimer's disease. Ann Neurol 40: 759–766

Masliah E, Raber J, Alford M, Mallory M, Mattson MP, Yang D, Wong D, Mucke L (1998) Amyloid protein precursor stimulates excitatory amino acid transport. Implications for roles in neuroprotection and pathogenesis. J Biol Chem 273: 12548–12554

Meyer T, Lenk U, Küther GL, Speer A, Weindl A, Ludolph AC (1995) Studies of the coding region of the neuronal glutamate transporter (EAAC1) gene in ALS. Ann Neurol 37: 817–819

Meyer T, Meyer B, Münch C, Sitte W, Küther G, Speer A, Ludolph AC (1996) The glial glutamate transporter cDNA in patients with amyotrophic lateral sclerosis. Ann Neurol 40: 456–459

Meyer T, Ludolph AC, Morkel M, Hagemeier C, Speer A (1997) Genomic organization of the human excitatory amino acid transporter gene GLT-1. Neuroreport 8: 775–777

Meyer T, Münch C, Knappenberger B, Liebau S, Völkel H, Ludolph AC (1998a) Alternative splicing of the glutamate transporter EAAT2. Neurosci Lett 241: 1–3

Meyer T, Münch C, Völkel H, Booms P, Ludolph AC (1998b) The EAAT 2 (GLT-1) gene in motor neuron disease: Absence of mutations in amyotrophic lateral sclerosis and a point mutation in individuals with hereditary spastic paraplegia. J Neurol Neurosurg Psychiatry, in press

Miller RG, Moore D, Young LA, Armon C, Barohn RJ, Bromberg MB, Bryan WW, Gelinas DF, Mendoza MC, Neville HE, Parry GJ, Petajan JH, Ravits JM, Ringel SP, Ross MA, the WALS Study Group (1996) Placebo-controlled trial of gabapentin in patients with amyotrophic lateral sclerosis. Neurology 47: 1383–1388

Mizoule J, Meldrum D, Mazadier M, Croucher M, Ollat C, Uzan A, Legrand J, Gueremy C, Le Fur G (1985) 2-amino-6-trifluoromethoxybenzothiazole, a possible antagonist of excitatory amino acid transmission. 1. Anticonvulsant properties. Neuropharmacol 24: 767–773

Münch C, Schwalenstöcker B, Liebau S, Völkel H, Ludolph AC, Meyer T (1998) 5′-Heterogeneity of the human glutamate transporter cDNA EAAT2 (GLT-1). Neuroreport 9: 1295–1297

Nagai M, Abe K, Okamoto K, Itoyama Y (1998) Identification of alternative splicing forms of GLT-1 mRNA in the spinal cord of amyotrophic lateral sclerosis patients. Neurosci Lett 244: 165–168

Ochs G, Penn RD, Beck M, Giess R, Magnus T, Sims T, Sendtner M, Toyka KV (1998) Intrathecal infusion of recombinant human-brain-derived neurotrophic factor (rhBDNF) is well tolerated in patients with ALS. Abstract, 9th International Symposium on ALS/MND

Olney JW (1969a) Brain lesions, obesity, and other disturbances in mice treated with monosodium glutamate. Science 164: 719–721

Olney JW (1969b) Glutamate-induced retinal degeneration in neonatal mice. Electron microscopy of the acutely evolving lesion. J Neuropathol Exp Neurol 28: 455–474

Parkes TL, Elia AJ, Dickinson D, Hilliker AJ, Phillips JP, Boulianne GL (1998) Extension of *Drosophila* lifespan by overexpression of human SOD1 in motorneurons. Nat Gen 19: 171–174

Penn RD, Kroin JS, York MM, Cedarbaum JM (1997) Intrathecal ciliary neurotrophic factor delivery for treatment of amyotrophic lateral sclerosis (phase I trial). Neurosurgery 40: 94–99

Perry TL, Hansen S, Jones K (1987) Brain glutamate deficiency in amyotrophic lateral sclerosis. Neurology 37: 1845–1848

Plaitakis A, Constantakakis E, Smith J (1988) The neuroexcitotoxic amino acids glutamate and aspartate are altered in the spinal cord and brain in amyotrophic lateral sclerosis. Ann Neurol 24: 446–449

Riepe M, Hori N, Ludolph AC, Carpenter DO, Spencer PS, Allen CA (1992) Inhibition of energy metabolism by 3-nitropropionic acid activates ATP-sensitive potassium channels. Brain Res 586: 61–66

Riepe M, Ludolph A, Seelig M, Spencer PS, Ludolph AC (1994) Increase of ATP levels by glutamate antagonists is unrelated to neuroprotection. Neuroreport 5: 2130–2132

Riepe HW, Hori N, Ludolph AC, Carpenter DO (1995) Failure of neuronal ion exchange, not potentiated excitation causes excitotoxicity after inhibition of oxidative phosphorylation. Neuroscience 64: 91–97

Riviere M, Meininger V, Zeisser P, Munsat T (1998) An analysis of extended survival in patients with amyotrophic lateral sclerosis treated with riluzole. Arch Neurol 55: 526–528

Rosen DR, Siddique T, Pattersson D, Figlewicz DA, Sapp P, Hentati A et al. (1993) Mutations in Cu/Zn superoxide dismutase gene are associated with familial amyotrophic lateral sclerosis. Nature 362: 59–62

Rothstein JD, Martin LJ, Kuncl RW (1992) Decreased brain and spinal cord glutamate transport in amyotrophic lateral sclerosis. N Engl J Med 326: 1464–1468

Rothstein JD, Van Kammen M, Levey AI, Martin L, Kuncl RW (1995) Selective loss of glial glutamate transporter GLT-1 in amyotrophic lateral sclerosis. Ann Neurol 38: 73–84

Sendtner M (1997) Gene therapy for motor neuron disease. Nat Med 3: 380–381

Shaw PJ, Forrest V, Ince PG, Richardson JP, Wastell HJ (1995) CSF and plasma amino acid levels in motor neuron disease: elevation of CSF glutamate in a subset of patients. Neurodegeneration 4: 209–216

Stefani A, Spadoni F, Bernardi G (1997) Differential inhibition by riluzole, lamotrigine, and phenytoin of sodium and calcium currents in cortical neurons: implications for neuroprotective strategies. Exp Neurol 147: 115–122

Taylor CP (1995) Gabapentin — mechanisms of action. In: Levy RH, Mattson RH, Meldrum BNM (eds) Antiepileptic drugs, 4th ed. Raven Press, New York, pp 829–841

Terro F, Lasort M, Vlader F, Ludolph A, Hugon J (1996) Antioxidant drugs block in vitro the neurotoxicity of CSF from patients with amyotrophic lateral sclerosis. Neuroreport 7: 1970–1972

Torp R, Lekieffre D, Levy LM, Haug FM, Danbolt NC, Meldrum BS, Ottersen OP (1995) Reduced postischemic expression of a glial glutamate transporter, GLT1, in the rat hippocampus. Exp Brain Res 103: 51–58

Tsai G, Stauch-Slusher B, Sim L et al. (1990) Reductions in acidic amino acids and N-acetyl-aspartyl-glutamate (NAAG) in amyotrophic lateral sclerosis CSF. Brain Res 556: 151–156

Wiedemann FR, Winkler K, Kuznetsov AV, Bartels C, Vielhaber S, Feistner H, Kunz WS (1998) Impairment of mitochondrial function in skeletal muscle of patients with amyotrophic lateral sclerosis. J Neurol Sci 156: 65–72

Wong PC, Pardo CA, Borchelt DR, Lee MK, Copeland NG, Jenkins NA, Sisodia SS, Cleveland DW, Price DL (1995) An adverse property of a familial ALS-linked SOD1 mutation causes motor neuron disease characterized by vacuolar degeneration of mitochondria. Neuron 14: 1105–1116

World Federation of Neurology Research Group on Neuromuscular Diseases (1994) El Escorial World Federation of Neurology criteria for the diagnosis of amyotrophic lateral sclerosis. J Neurol Sci 124(Suppl.): 96–107

Authors' address: Dr. A. C. Ludolph, Department of Neurology, University of Ulm, Steinhövelstrasse 9, D-89075 Ulm, Federal Republic of Germany

Antiglutamate therapies in Huntington's disease

K. Kieburtz

University of Rochester, Rochester, New York, U.S.A.

Summary. Huntington's disease is an autosomal dominant neurodegenerative disorder caused by an unstable trinucleotide CAG repeat. The mechanism by which the genetic defect leads to neuronal injury and death is unknown, but is thought to include glutamate-mediated excitotoxicity and abnormalities of mitochondrial energy production. Both of these mechanisms may lead to a final common pathway of increased production of free radical species. Prior clinical trials in patients with Huntington's disease that have addressed these hypotheses have been limited by size. A current, NIH-funded trial of remacemide hydrochloride and Coenzyme Q10 in 340 patients with Huntington's disease is described. This is the largest and longest multi-center trial in Huntington's disease to address the glutamate- and mitochondrial-mediated hypotheses of neurodegeneration.

Introduction

Huntington's disease (HD) is an autosomal dominant, fully penetrant neurodegenerative disorder which results from an unstable expansion of the trinucleotide CAG repeat in the IT15 gene located near the telomere of the short arm of chromosome 4 (Huntington's Disease Collaborative Research Group, 1993). The clinical features of HD usually emerge in adulthood with disorders of voluntary movement and chorea. The disease presents with progressive deterioration of cognitive function, behavior and motor control, eventually leading to functional disability and death over a period of approximately 25 years (Greenamyre and Shoulson, 1994). The pathology of HD is characterized by selective degeneration of spiny neurons in the caudate and putamen as well as diffuse brain atrophy. Neuronal loss in the striatum is largely confined to GABAergic, medium-sized, spiny striatal neurons which project to the globus pallidus and which receive glutamatergic inputs from cerebral cortex (Albin et al., 1989). Aspiny interneurons are relatively spared.

A central question in HD research is how an abnormal gene product that is expressed throughout the body and brain produces focal and selective neuropathology. Neurotoxicologic approaches have been used to develop animal models which show characteristics resembling the focal pathology of

HD. These models are derived from the independent and additive effects of NMDA glutamate receptor-mediated glutamatergic neurotransmission (excitotoxicity) and mitochondrial energy defects leading to increased production of free radicals. The combined effects result in neural injury and degeneration. The first animal paradigm of HD was produced by injections of the excitotoxin kainic acid into the striatum of rats (Coyle and Schwarcz, 1976). Later models have been produced by injecting selective NMDA receptor agonists (e.g., quinolinic acid) that destroy striatal spiny projection neurons while sparing other interneurons (Schwarcz et al., 1982; Beal et al., 1986; Beal et al., 1991). Spiny neurons in the striatum have more NMDA receptors than striatal interneurons (Landwehrmeyer et al., 1994). In the HD brain, NMDA receptor levels are decreased early in the illness, consistent with the hypothesis that neurons with high levels of NMDA receptors are especially vulnerable to injury and degeneration (Graveland et al., 1985; Young et al., 1988; Albin et al., 1990).

Excitatory glutamatergic transmission may interact with cellular energy impairment, by increasing cellular energy demands and causing calcium ions to flow into neurons (Choi, 1988). The interrelationships between energy metabolism and glutamate receptor activation have recently been emphasized in animal models of HD produced by mitochondrial inhibitors. Malonate (Greene et al., 1993) and 3-nitroproprionic acid (Beal et al., 1993a) produce selective damage in spiny neurons when they are injected into the striatum. NMDA receptor antagonists as well as mitochondrial energy buttressing agents can attenuate striatal lesions induced by injections of malonate (Beal et al., 1993a; Beal et al., 1994; Beal et al., 1993b).

Based on the similarity of selective striatal pathology in excitotoxic animal models of HD and the efficacy of glutamate receptor antagonists and mitochondrial buttressing agents in limiting the neuronal injury, we reasoned that it was rational to pursue a clinical trial of an NMDA channel blocker and an agent that enhances mitochondrial metabolism in patients who are in the early stages of HD.

Remacemide hydrochloride is a non-competitive NMDA receptor antagonist acting at the ion channel site of the receptor complex. In an autoradiographic binding assay it has an IC50 of about $800\,\mu M$ (Porter and Greenamyre, 1995). Remacemide hydrochloride is unusual in that it is an orally available glutamate receptor antagonist and has been generally well tolerated in prior clinical research (Kieburtz et al., 1996). Prior clinical trials exploring the anti-glutamate rationale have been limited by the lack of potent, specific and tolerable interventions, as well as by an insufficient sample size to detect meaningful therapeutic effects. Baclofen and lamotrigine are agents which attenuate glutamate neurotransmission in part by inhibiting glutamate release. A placebo-controlled, randomized trial of baclofen in HD did not favorably influence the rate of illness progression (Shoulson et al., 1989b). This study involved 60 randomized patients and had sufficient power to detect only very large therapeutic effects. A randomized, placebo-controlled trial of lamotrigine in 60 patients at the University of British Columbia also found no benefit of the intervention in slowing the progression of illness (Kremer et al., 1997). Both studies were limited by a small sample size. An initial tolerability trial has been conducted with remacemide in HD patients (Kieburtz et al., 1996). In this randomized, parallel group, double-blind, placebo-controlled dosage-ranging trial 31 patients were randomized to 3 arms (placebo, remacemide 200 mg/day and remacemide 600 mg/day). No significant clinical improvements were observed or expected over 6 weeks of treatment, but the intervention was well tolerated. Combining these observations as well as the lessons learned from prior experience in designing clinical trials in HD, we have commenced a trial of remacemide hydrochloride and the mitochondrial buttressing agent Co-enzyme Q10 (CoQ) in patients with HD.

The Co-enzyme Q10 and Remacemide Evaluation in Huntington's Disease (CARE-HD) study plans to evaluate 340 ambulatory patients with HD enrolled at 23 investigative sites in North America. Eligible subjects were those with HD as defined by the characteristic movement disorder in the setting of a confirmatory CAG expansion, and who were in Stage I or II of illness (Shoulson et al., 1989a) [total functional capacity (TFC) ≥ 7] and who were 14 years of age or older. Eligible subjects wee randomized to 1 of 4 treatment arms using a 2×2 factorial design (placebo, Co-Q alone, remacemide alone and the combination of Co-Q and remacemide). The 2×2 factorial design allows us to assess the independent and combined efficacy of remacemide and Co-Q in slowing the progression of HD. After randomization patients will be followed prospectively and systematically for 30 months. The TFC scale has been used as a primary outcome measure in HD clinical trials as well as in large natural history data bases (Shoulson et al., 1996). The TFC has been validated against radiographic measures of disease progression including CT and MR and fluorodeoxyglucose positron emission tomography measures of striatal metabolism (Bamford et al., 1989; Young et al., 1986a; Young et al., 1986b). In an inter-rater reliability study the TFC has been found to be reliable and used by a variety of health professionals (Shoulson et al., 1989a). The TFC shows a steady linear decline over the course of illness and is especially sensitive to changes in the early and mid-stages of disease which is the focus of the proposed trial (Young et al., 1986b; Feigin et al., 1995; Myers et al., 1991; Penney et al., 1990). After screening and baseline assessments, follow up visits will be conducted at 1, 4, 8, 12, 16, 20, 25 and 30 months after baseline. The Unified Huntington's Disease Rating Scale (UHDRS), which incorporates the TFC scale as well as other measures of motor and cognitive function, will be administered at each visit (Huntington Study Group, 1996). Routine safety laboratory measures will be obtained at the 1, 8, 20 and 30 month visits. Serum blood concentrations of remacemide and Co-Q will be measured at the 8 and 30 month visit. At the baseline and final study visit on experimental medications additional supplemental neuropsychologic testing will be administered.

The primary outcome measure will be the change in total functional capacity between the baseline and 30 month visit. We considered performing TFC ratings on two separate occasions at both baseline and 30 months in hopes of reducing the variability of our primary outcome measure; however, data from the 4 week safety and tolerability trial of remacemide indicated that the average of 2 ratings performed by the same rater 2 or 4 weeks apart is essentially no different from a single rating. The secondary measures will include several clinical efficacy measures derived from the UHDRS and the supplemental neuropsychologic test battery including the Hopkins Verbal Learning Test (Brandt, 1991), the brief test of attention (Schretlen et al., 1996), the Trail Making Test (Reitan and Woflson, 1958) and the Conditional Associative Learning Test (Petrides, 1990). The primary statistical analysis will involve an analysis of covariance model with treatment as the factor of interest, investigator as a stratification factor and TFC at baseline as a covariant. F tests will be performed for significance of the main effects of Co-Q and remacemide, and confidence intervals will be constructed for each of these effects. Other pairwise comparisons among the 4 treatment combinations will be performed in a similar manner but will be considered of secondary importance. A decision to use the change in TFC score from baseline to 30 months as the primary analysis was made after considering and rejecting several analytic strategies including individual slopes of decline, that presume disease progression is linear in all treatment groups. Based on this analysis plan and our preliminary data, we have calculated that we have 80% power to detect a 40% attenuation of clinical decline.

Recruitment into the trial was virtually completed after 6 months in large part due to intense enthusiasm among HD patients and families, and HD investigators. Once all subjects are enrolled, follow up clinical evaluations and finalization of the database will take an additional 3 years before results are available.

Acknowledgement

Supported in part by PHS NS ROI NS 35 284-02.

References

Albin RL, Young AB, Penney JB (1989) The functional anatomy of basal ganglia disorders. Trends Neurosci 12: 366–375

Albin RL, Young AB, Penney JB, Handelin B, Balfour R, Anderson KD, Markel DS, Tourtellotte WW, Reiner A (1990) Abnormalities of striatal projection neurons and N-methyl-D-aspartate receptors in presymptornatic Huntington's disease. N Engl J Med 332: 1293–1298

Bamford KA, Caine ED, Kido DR, Plassche WM, Shoulson I (1989) Clinical-pathologic correlation in Huntington's disease: A neuropsychological and computed tomography study. Neurology 39: 796–801

Beal MF, Kowall NW, Ellison DW, Mazurek MF, Swartz KJ, Martin JB (1986) Replication of the neurochemical characteristics of Huntington's disease by quinolinic acid. Nature 321: 168–171

Beal MF, Ferrante RJ, Swartz KJ, Kowall NW (1991) Chronic quinolinic acid lesions in rats closely resemble Huntington's disease. J Neurosci 11: 1649–1659

Beal W, Brouillet E, Jenkins BG, Ferrante RJ, Kowall NW, Miller JM, Storey E, Srivastava R, Rosen BR, Hyman BT (1993a) Neurochemical and histological characterization of striatal excitotoxic lesions produced by the mitochondrial toxin 3-nitropropionic acid. J Neurosci 13: 4181–4192

Beal MF, Brouillet E, Jenkins BG, Henshaw R, Rosen B, Hyman BT (1993b) Age-dependent striatal excitotoxic lesions produced by the endogenous mitochondrial inhibitor malonate. J Neurochem 61: 1147–1150

Beal MF, Henshaw DR, Jenkins BG, Rosen BR, Schulz JB (1994) Coenzyme Q10 and nicotinamide block striatal lesions produced by the mitochondrial toxin malonate. Ann Neurol 36: 882–888

Brandt J (1991) The Hopkins verbal learning test: development of a new memory test with six equivalent forms. Clin Neuropsychol 5: 125–142

Choi DW (1988) Glutamate neurotoxicity and diseases of the nervous system. Neuron 1: 623–634

Coyle JT, Schwarcz R (1976) Lesion of striatal neurons with kainic acid provides a model for Huntington's chorea. Nature 263: 244–246

Feigin A, Kieburtz K, Bordwell K, Como P, Steinberg K, Sotack J, Zimmerman C, Hickey C, Orme C, Shoulson I (1995) Functional decline in Huntington's disease. Mov Disord 10: 211–214

Graveland GA, Williams RS, DiFiglia M (1985) Evidence of degenerative and regenerative changes in neostriatal spiny neurons in Huntington's disease. Science 227: 770–773

Greenamyre JT, Shoulson I (1994) Huntington's Disease. In: Calne DB (ed) Neurodegenerative disease. Saunders, Philadelphia, pp 685–704

Greene JG, Porter MP, Eller RV, Greenamyre JT (1993) Inhibition of succinate dehydrogenase by malonic acid produces an "excitotoc" lesion in rat striatum. J Neurochem 61: 1151–1154

The Huntington's Disease Collaborative Research Group (1993) A novel gene containing a trinucleotide repeat that is expanded and unstable on Huntington's disease chromosomes. Cell 72: 971–983

The Huntington Study Group (1996) Unified Huntington's Disease Rating Scale: Reiliability and Consistency. Mov Disord 11: 136–142

Kieburtz K, Feigin A, McDermott M, Como P, Abwender D, Zimmerman C, Hickey C, Orme C, Claude K, Sotack J, Greenamyre JT, Dunn C, Shoulson I (1996) A controlled trial of the glutamate antagonist remacemide hydrochloride in Huntington's disease. Mov Disord 11: 273–277

Kremer B, Clark CM, Hardy M, Almqvist E, Raymond L, Hayden MR (1997) Lamotrigine does not retard the progression of Huntington's disease. 17th International Meeting of the World Federation of Neurology Research Group on Huntington's Disease 8/30–9/2/97 (abstract)

Landwehrmeyer GB, Standaert DG, Testa CM, Penney JB, Young AB (1994) NMDA receptor subunit expression by rat striatal projection neurons and intemeurons. J Neurosci 14: 5297–5308

Myers RH, Sax DS, Koroshetz WJ, Mastromauro C, Cupples LA, Kiely DK, Pettengill FK, Bird ED (1991) Factors associated with slow progression in Huntington's disease. Arch Neurol 48: 800–804

Penney JB Jr, Young AB, Shoulson I, Starosta-Rubenstein S, Snodgrass SR, Sanchez Ramos J, Ramos-Arroyo M, Gomez F, Penchaszadeh G, Alvir J, Esteves J, DeQuiroz I, Marsol N, Moreno H, Conneally PM, Bonilla E, Wexler NS (1990) Huntington's disease in Venezuela: 7 years of follow-up on symptomatic and asymptomatic individuals. Mov Disord 5: 93–99

Petrides M (1990) Nonspatial conditional learning impaired in patients with unilateral frontal but not unilateral temporal lobe excisions. Neuropsychologia 28: 137–149

Porter RHP, Greenamyre JT (1995) Regional variations in the pharmacology of NMDA receptor channel blockers: implications for therapeutic potential. J Neurochem 64: 614–623

Reitan RM, Wolfson D (1958) The Halstead-Reitan Neuropsychological Test Batttery. Neuropsychology Press, Tucson

Schretlen D, Bobholz JH, Brandt J (1996) Development and psychometric properties of the Brief Test of Attention. Clin Neuropsychol 10: 80–89

Schwarcz R, Whetsell WO, Mangano RM (1982) Quinolinic acid: An endogenous metabolite that produces axon-sparing lesions in rat brain. Science 219: 316–318

Shoulson I, Kurlan R, Rubin AJ, Goldblatt D, Behr J, Miller C, Kennedy J, Bamford KA, Caine ED, Kido DK, Plumb S, Odoroff C (1989a) Assessment of functional capacity in neurodegenerative movement disorders: Huntington's disease as a prototype. In: Munsat TL (ed) Quantification of neurological deficit. Butterworths, Boston, pp 271–283

Shoulson I, Odoroff C, Oakes D, Behr J, Goldblatt D, Caine E, Kennedy J, Miller C, Bamford K, Rubin A, Plumb S, Kurlan R (1989b) A controlled clinical trial of baclofen as protective therapy in early Huntington's disease. Ann Neurol 25: 252–259

Young AB, Penney JB, Starosta-Rubinstein S, Markel DS, Berent S, Giordani B, Ehrenkaufer R, Jewett D, Hichwa R (1986a) PET scan investigations of Huntington's disease: cerebral metabolic correlates of neurological features and functional decline. Ann Neurol 20: 296–303

Young AB, Shoulson I, Penney JB, Starosta-Rubinstein S, Gomez F, Travers H, Ramos M, Snodgrass SR, Bonilla A, Moreno H, Wexler N (1986b) Huntington's disease in Venezuela: neurological features and functional decline. Neurology 36: 244–249

Young AB, Greenamyre JT, Hollingsworth Z, Albin R, D'Amato C, Shoulson I, Penney JB (1988) NMDA receptor losses in putamen from patients with Huntington's disease. Science 241: 981–983

Author's address: K. Kieburtz, MD, University of Rochester, 1351 Mt. Hope Avenue, Suite 220, Rochester, New York, U.S.A.

Neural transplantation in animal models of multiple system atrophy: a review

G. K. Wenning, R. Granata, Z. Puschban, C. Scherfler, and **W. Poewe**

Neurological Research Laboratory, Department of Neurology, University Hospital,
Innsbruck, Austria

Summary. Multiple system atrophy of the striatonigral degeneration (MSA-SND) type is increasingly recognized as major cause of neurodegenerative parkinsonism. Due to combined degeneration of substantia nigra pars compacta (SNC) and of striatum, antiparkinsonian therapy based on levodopa substitution eventually fails in more than 90% of patients. Animal models of MSA-SND are urgently required as test-bed for the evaluation of novel therapeutic interventions in this disorder such as neurotrophic factor delivery and neuronal transplantation. A number of well established rodent and primate models of Parkinson's (PD) and Huntington's (HD) disease replicate either nigral ("PD-like") or striatal ("HD-like") pathology and may therefore provide a useful baseline for the development of MSA-SND models. Previous attempts to mimick MSA-SND pathology in rodents have included sequential injections of 6-hydroxydopamine (6OHDA) and quinolinic acid (QA) into medial forebrain bundle and ipsilateral striatum, respectively ("double toxin — double lesion" approach). Preliminary evidence in rodents subjected to such lesions indicates that embryonic transplantation may partially reverse behavioural abnormalities. Intrastriatal injections of mitochondrial toxins such as 3-nitropropionic acid (3NP) and 1-methyl-4-phenylpyridinium (MPP+) in rodents result in (secondary) excitotoxic striatal lesions and subtotal neuronal degeneration of ipsilateral SNC, thus producing MSA-SND-like pathology by a simplified "single toxin — double lesion" approach. Comparative studies of human SND pathology and rodent striatonigral lesions are required in order to determine the rodent model(s) most closely mimicking the human disease process.

Introduction

Multisystem neurodegenerative diseases run a relentlessly progressive course that is poorly understood and uninfluenced by all medical and surgical measures. In most of these disorders the neuropathological lesion pattern is dominated by involvement of two or more neuronal systems. In multiple system atrophy (MSA) of the striatonigral degeneration (SND) type two related neuronal systems, substantia nigra and striatum, that have been studied in great detail experimentally, bear the brunt of neuronal degeneration (Fearnley and Lees, 1990; Tison et al., 1995). After Parkinson's disease (PD) MSA-SND represents the most common neurodegenerative disorder accounting for parkinsonism (Wenning and Quinn, 1997b). Due to a poor response to antiparkinsonian therapy resulting from striatal degeneration with loss of dopamine receptors, disease progression is more rapid in MSA-SND compared to PD. Indeed, median survival in a recent clinical series was 9.3 years (Wenning et al., 1994). Other clinical features include autonomic failure, cerebellar and/or pyramidal signs (Quinn, 1994).

Neuropathologically, other regions may also be affected to a lesser degree in MSA-SND including inferior olives, pontine nuclei and cerebellar Purkinje cells (i.e. olivopontocerebellar atrophy) (Wenning et al., 1996c). It is presently unknown why certain areas of the brain are affected while others are spared in MSA.

This disorder appears highly suitable for the development of animal models 1) because the neuropathology of the disease is known to a large extent and 2) because animal models of isolated nigral or striatal lesions have been established for many years, and can serve as a framework for the development of MSA-SND animal models. At present there is no universally accepted animal model of MSA-SND except for one preliminary attempt (Wenning et al., 1996a). Deutch et al. (Deutch et al., 1989) proposed an animal model of olivopontocerebellar atrophy-associated parkinsonism using systemic administration of 3-acetylpyridine (3-AP) in rats. However, loss of nigrostriatal dopamine projections was incomplete without definite neuronal depletion in substantia nigra pars compacta and striatal neurons appeared preserved.

Neuropathology of SND

In contrast to Huntington's disease (HD) the putamen bears the brunt of striatal pathology the caudate nucleus being less severely involved. There is a distinct topographical distribution with a predilection for the posterior two-thirds and dorsolateral putamen (Fearnley et al., 1990). The medium-sized GABA-ergic spiny neurons appear to be particularly affected as shown by calcineurin immunocytochemistry (Daniel, 1992). Both direct and indirect striatal projections are involved as shown by loss of substance P (SP) and met-enkephalin (Menk) immunoreactivity. Furthermore, the reciprocal relationship between substantia nigra and putamen appears to be damaged (Goto et al., 1989a,b). Biochemical studies suggest a marked loss of striatal neurotransmitters such as dopamine and noradrenaline (Spokes et al., 1979). Regional analysis of cholineacetyltransferase (Chat) activity shows a reduction in the striatum of some, but not all MSA cases (Spokes et al., 1979). Markers of aspiny striatal neurons such as nicotinamide adenine dinucleotide phosphatate dehydrogenase (NADPH-d), somatostatin and neuropeptide Y remain to be determined. Furthermore, mRNA levels of specific nigrostriatal markers such as tyrosine hydroxylase (TH), SP and Menk have not been reported.

Characteristic oligodendroglial cytoplasmic inclusions (GCIs) have recently been identified in MSA brains (Papp et al., 1989; Lantos, 1997). Subsequently, inclusions have also been reported in the neuronal cytoplasm and both neuronal and glial nuclei (Papp and Lantos, 1992; Kato and Nakamura, 1990). The pattern of GCI deposition is remarkably selective for the cortico-striato-cortical motor loop (Papp and Lantos, 1994). They are found in the putamen and caudate nucleus, but also frequently in frontal projection areas unaffected by significant cell loss and gliosis. The discovery of GCIs in MSA brain areas not exhibiting neuronal cell loss suggests that the disease may primarily affect oligodendroglia. Consistent with this assumption, recent postmortem studies revealed evidence of oligodendroglial, but not neuronal, apoptosis in MSA brains (Probst-Cousin et al., 1998). Indeed, MSA may represent a unique oligodendroglial disease resulting in a secondary neuronal multisystem degeneration (Lantos, 1997).

Animal models of MSA-SND

Double toxin — double lesion approach

We have recently developed a rodent model of MSA-SND based on a sequential stereotaxic injection of a nigral and a striatal neurotoxin (Wenning et al., 1996a; Wenning et al., 1996b). 6-Hydroxydopamine (6-OHDA) was administered into the left medial forebrain bundle of male Wistar rats, followed 3 to 4 weeks later by intrastriatal injection of quinolinic acid (QA) into the ipsilateral striatum. The 6-OHDA lesion resulted in ipsiversive rotation to amphetamine and contraversive rotation to apomorphine. Following the subsequent striatal lesion, amphetamine-induced ipsilateral rotation persisted, but apomorphine-induced contralateral rotation was reduced or abolished.

Single toxin — double lesion approach

A well described animal model of PD is produced by the systemic injection of 1-methyl-4-phenyl-1,2,3,6-tetrahydropyridine (MPTP) into mice or primates (Bloem et al., 1990). MPTP, which is metabolized to the mitochondrial complex I inhibitor 1-methyl-4-phenylpyridinium (MPP+), causes dopaminergic cell death in the substantia nigra pars compacta (SNC), presumably due to mitochondrial impairment (Sonsalla et al., 1992). In the rat systemic administration of MPTP is not effective due to lack of conversion into the active agent MPP+. However, direct intrastriatal injection causes, in addition to a considerable local lesion, profound loss of dopaminergic SNC neurons (Frim et al.,

1994). The striatal lesion appears to be excitotoxic in nature (Storey et al., 1992). The PD toxin MPP+ may therefore be converted into a potential SND toxin by changing the systemic mode of administration to local striatal injection, producing a novel double lesion model that mimicks neuronal degeneration in two related neuronal systems. Further studies are required to examine the degree and time course of nigral versus striatal degeneration obtained with intrastriatal MPP+ injections. Striatal injection of 6-OHDA results in progressive neuronal cell loss in substantia nigra pars compacta reaching about 15–20% of normal 16 weeks after administration (Sauer and Oertel, 1994). It may be postulated that intrastriatal injection of MPP+ also produces such a chronic model of parkinsonism. The dopaminergic neurotoxicity of both 6-OHDA and MPP+ may largely depend on their incorporation into dopaminergic neurons via uptake by the dopamine transporter (Javitch et al., 1985; Sundstrom et al., 1986; Pifl et al., 1993).

Rodent models of HD based on intrastriatal injections of n-methyl-D-aspartate (NMDA) agonists such as QA or ibotenic acid (Huang et al., 1995) and mitochondrial respiratory chain (MRC) toxins such as malonate and 3-nitropropionic acid (3-NP) (Wüllner et al., 1994) have been studied in much detail. The neuropathological damage that is provoked by intrastriatal injections of excitotoxic and MRC agents appears to replicate some of the salient neurochemical and immunohistochemical changes observed in the human disease (Beal et al., 1993; Beal et al., 1986; Beal et al., 1989). The rodent models facilitated the development of primate models (Hantraye et al., 1990) and also served to evaluate therapeutic intervention (Emerich et al., 1997). Recently, preliminary evidence has been reported to suggest that some HD toxins may be helpful in developing a SND model.

3-acetylpyridine (3-AP), when given systemically to rodents, exerts a neurotoxic effect on nigrostriatal dopaminergic neurons without attacking the mesolimbic dopamine system or striatal neurons (Deutch et al., 1989; Takada et al., 1993). Recent studies have demonstrated that 3-AP induces an active type of cell death with ultrastructural features distinct from apoptosis (Wüllner et al., 1997). Intrastriatal injections of 3-AP produce an excitotoxic-like striatal lesion (Schulz et al., 1994) and may also cause retrograde nigral degeneration taking into account the previously shown striatal dopamine depletion. This postulated nigral toxicity of 3-AP when injected into striatum remains to be demonstrated.

In contrast to 3-AP, systemic administration of 3-NP produces excitotoxic-like lesions regionally restricted to the striatum in both rats (Beal et al., 1993) and non-human primates (Brouillet and Hantraye, 1995) sparing the SNC neurons. Interestingly, intrastriatal injections of 3-NP were recently reported to deplete TH positive nigral neurons in addition to striatal (Nakao and Brundi, 1997) suggesting that this neurotoxin may lesion both substantia nigra and striatum by altering the systemic mode of administration to a local intrastriatal injection. Preliminary observations in our laboratory confirm the reported 3-NP induced nigral lesion with approximately 50% loss of dopaminergic neurons (Fig. 1a,b).

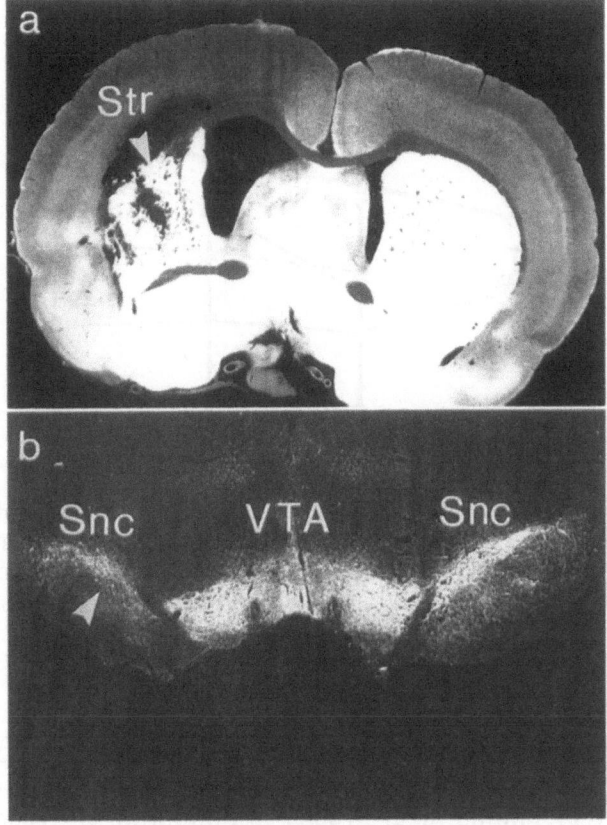

Fig. 1. a Photomicrograph of acetylcholinesterase stained section of a 3-nitropropionic acid lesioned striatum (*Str*). **b** Depletion of tyrosinehydroxylase immunoreactivity in ipsilateral substantia nigra pars compacta (*Snc*), (*VTA* ventral tegmental area). Images were obtained using immunostained slices as negatives in a photographic enlarger. Thus immunoreactive structures appear white

Therapeutic intervention in MSA-SND animal models

In recent years, gene therapy has been actively investigated as potential treatment for PD or HD. Most studies have followed an ex vivo approach. A neuroprotective effect of brain derived neurotrophic factor (BDNF) (Galpern et al., 1996) and glial cell line derived neurotrophic factor (GDNF) (Winkler et al., 1996; Gash et al., 1996) in animal models of PD and of neurotrophic factor (NGF) and ciliary neurotrophic factor (CNTF) in animal models of HD (Martinez-Serrano and Björklund, 1996; Anderson et al., 1996; Emerich et al., 1997) has been demonstrated. However, several important issues, such as maintenance of long-term high-level transgene expression and its differential effect on target cell populations remain unresolved. At the same time, transplantation of embryonic CNS tissue has been firmly established in PD and HD animal models (Olanow et al., 1996; Peschanski et al., 1996) and longterm clinical trials have demonstrated lasting benefits in PD (Wenning et al., 1997a). So far, no studies have been reported on the effect of neurotrophic

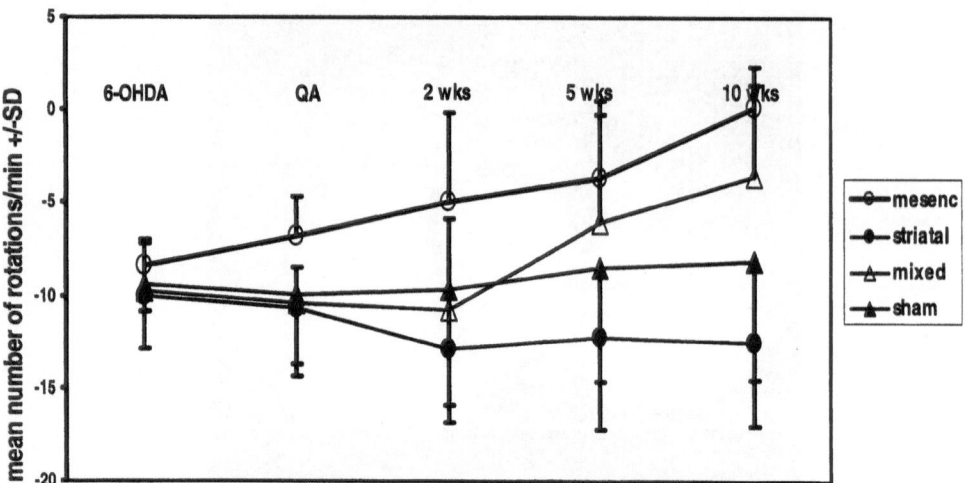

Fig. 2. Mean amphetamine-induced rotations in grafted or sham animals following 6-OHDA and QA lesions as well as 2, 5 and 10 weeks after transplantation. Rotation behaviour was recorded for 60 minutes after i.p. injection of 5 mg/kg amphetaminesulfate. *Mesenc* mesencephalic graft, *striatal* striatal graft, *mixed* mesencephalic-striatal cograft, *sham* saline injection

factors or embryonic transplants in MSA-SND animal models except one preliminary rodent study showing partial functional benefits induced by embryonic grafts (Wenning et al., 1996a; Wenning et al., 1996b). Prior to transplantation these animals had received a sequential lesion of unilateral SNC and striatum as outlined above. Subsequently, the lesioned striatum was implanted with fetal CNS allografts consisting of cell suspensions derived from both striatal primordium and ventral mesencephalon, or either alone.

Rats receiving a mixed or pure mesencephalic graft showed a significant reduction of amphetamine-induced rotation 10 weeks following transplantation compared to QA administration (mixed graft $p < 0,005$; mesencephalic graft: $p < 0,002$). This was not observed in animals receiving striatal grafts alone (Fig. 2). Apomorphine-induced contraversive rotation was significantly increased 10 weeks following striatal, but not mesencephalic, mixed or sham, transplantation compared to QA administration ($p < 0,05$) (Fig. 3) Wenning et al., 1996a; Wenning et al., 1996b). The characteristic drug-induced rotational response associated with the sequential nigral and striatal lesion was therefore partly reversed by pure mesencephalic and mixed mesencephalic-striatal grafts. TH and dopamine and cyclic adenosine 3′:5′-monophosphate-regulated phosphoprotein 32 (DARPP 32) immunocytochemistry showed mesencephalic and striatal graft survival in most animals. There was no significant difference in the number of surviving dopaminergic neurons in rats receiving pure mesencephalic versus mixed grafts (Fig. 4). The additional striatal graft component therefore failed to enhance the behavioural recovery and survival of grafted dopaminergic neurons in our double-lesion MSA-SND model. This is in contrast to the reported amelioration of rotational recovery

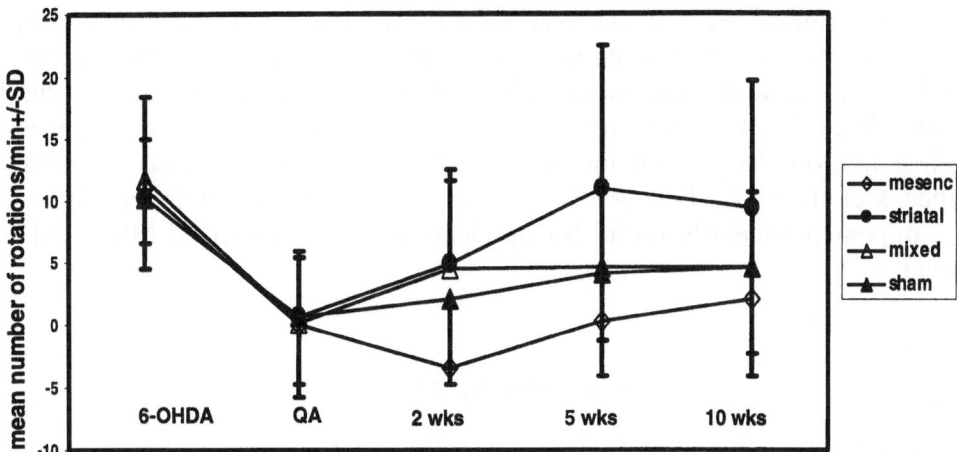

Fig. 3. Mean apomorphine-induced rotations in grafted or sham animals following 6-OHDA and QA lesions as well as 2, 5 and 10 weeks after transplantation. Rotation behaviour was recorded for 30 minutes after s.c. injection of 0,5 mg/kg apomorphine-hydrochloride. *Mesenc* mesencephalic graft, *striatal* striatal graft, *mixed* mesencephalic-striatal cograft, *sham* saline injection

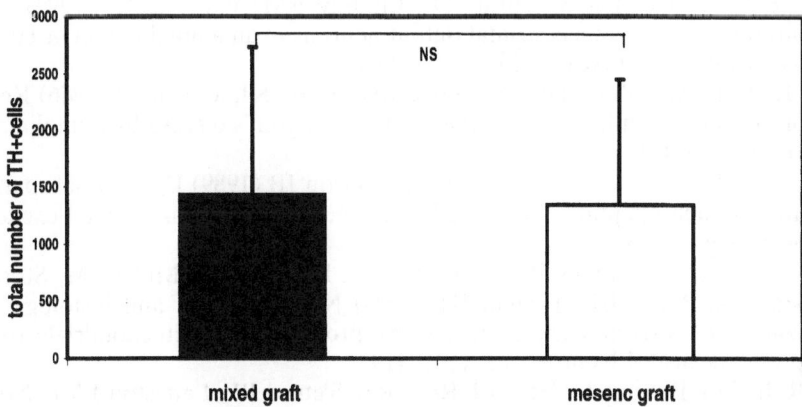

Fig. 4. Tyrosinehydroxylaseimmunohistochemistry of grafts in the MSA rat model. There was no significant difference between mean total number of TH positive cells between animals receiving mixed grafts and those receiving pure mesencephalic grafts. *Mesenc* mesencephalic graft, *mixed* mesencephalic-striatal cograft, *NS* non-significant

induced by embryonic dopaminergic neurons cografted with striatal cell suspensions in PD rat models (Brundin et al., 1986; Costantini and Snyder-Keller, 1997; Costantini et al., 1994; Yurek et al., 1990).

Future direction

In future studies the double lesion rodent MSA-SND model needs to be further optimized by matching of human and rodent neuropathology. This optimized model can then be used to improve the efficacy of embryonic transplantation. Furthermore, neuroprotective strategies including neurotrophic factor gene therapy may also be evaluated. A number of neurotrophic factors have been shown to protect both striatal and nigral neurons against neurotoxicity insult including GDNF (Perez-Navarro et al., 1996; Lapchak et al., 1997; Choi-Lundberg et al., 1997). Such trophic factors therefore appear particularly useful for application in the optimized MSA-SND model.

Acknowledgement

We thank Professor Dr Niall Quinn for critical reading of the manuscript. This work has been supported by the Parkinson's Disease Foundation, U.K., and the Austrian Science Foundation.

References

Anderson KD, Panayotatos N, Coran TL, Lindsay RM, Wiegand SJ (1996) Ciliary neurotrophic factor protects striatal output neurons in an animal model of Huntington disease. Proc Natl Acad Sci USA 93: 7346–7351

Beal MF, Kowall NW, Ellison DW, Mazurek MF, Swartz KJ, Martin JB (1986) Replication of the neurochemical characteristics of Huntington's disease by quinolinic acid. Nature 321: 168–171

Beal MF, Kowall NW, Swartz KJ, Ferrante RJ, Martin JB (1989) Differential sparing of somatostatin-neuropeptide Y and cholinergic neurons following striatal excitotoxic lesions. Synapse 3: 38–47

Beal MF, Brouillet E, Jenkins BG, Ferrante RJ, Kowall NW, Miller JM, Storey E, Srivastava R, Rosen BR, Hyman BT (1993) Neurochemical and histologic characterization of striatal excitotoxic lesions produced by mitochondrial toxin 3-nitropropionic acid. J Neurosci 13: 4181–4192

Bloem BR, Irwin I, Buruma OJ, Haan J, Roos RA, Tetrud JW, Langston JW (1990) The MPTP model: versatile contributions to the treatment of idiopathic Parkinson's disease. J Neurol Sci 97: 273–293

Brouillet E, Hantraye P (1995) Effects of chronic MPTP and 3-nitropropionic acid in nonhuman primates. Curr Opin Neurol 8: 469–473

Brundin P, Isacson O, Gage FH, Björklund A (1986) Intrastriatal grafting of dopamine-containing neuronal cell suspensions: effects of mixing with target or non-target cells. Brain Res 389: 77–84

Choi-Lundberg DL, Lin Q, Chang YN, Chiang YL, Hay CM, Mohajeri H, Davidson BL, Boh MC (1997) Dopaminergic neurons protected from degeneration by GDNF gene therapy. Science 275: 838–841

Costantini LC, Snyder-Keller A (1997) Co-transplantation of fetal lateral ganglionic eminence and ventral mesencephalon can augment function and development of intrastriatal transplants. Exp Neurol 145: 214–227

Costantini LC, Vozza BM, Snyder-Keller AM (1994) Enhanced efficacy of nigral-striatal cotransplants in bilaterally dopamine-depleted rats: an anatomical and behavioral analysis. Exp Neurol 127: 219–231

Daniel SE (1992) Multiple system atrophy. In: Bannister R, Mathias C (eds) Autonomic failure — a textbook of clinical disorders of autonomic nervous system. Oxford Medical Publications, Oxford, pp 564–585

Deutch AY, Rosin DL, Goldstein M, Roth RH (1989) 3-Acetylpyridine-induced degeneration of the nigrostriatal dopamine system: an animal model of olivoponto-cerebellar atrophy-associated parkinsonism. Exp Neurol 105: 1–9

Emerich DW, Winn SR, Hantraye PM, Peschanski M, Chen EY, Chu Y, McDermott P, Baetge EE, Kordower JH (1997) Protective effect of encapsulated cells producing neurotrophic factor CNTF in a monkey model of Huntington's disease. Nature 386: 395–399

Fearnley JM, Lees AJ (1990) Striatonigral degeneration. A clinicopathological study. Brain 113: 1823–1842

Frim DM, Uhler TA, Galpern WR, Beal MF, Breakefield XO, Isacson O (1994) Implanted fibroblasts genetically engineered to produce brain-derived neurotrophic factor prevent 1-methyl-4-phenylpyridium toxicity to dopaminergic neurons in the rat. Proc Natl Acad Sci USA 91: 5104–5108

Galpern WR, Frim DM, Tatter SB, Altar CA, Beal MF, Isacson O (1996) Cell-mediated delivery of brain-derived neurotrophic factor enhances dopamine levels in an MPP+ rat model of substantia nigra degeneration. Cell Transplant 5: 225–232

Gash DM, Zhang Z, Ovadi A, Cass WA, Yi A, Simmerman L, Russell D, Martin D, Lapchak PA, Collins F, Hoffer BJ, Gerhardt GA (1996) Functional recovery in parkinsonian monkeys treated with GDNF. Nature 380: 252–255

Goto S, Hirano A, Matsumoto S (1989a) Subdivisional involvment of nigrostriatal loop in idiopathic Parkinson's disease and striatonigral degeneration. Ann Neurol 26: 766–770

Goto S, Hirano A, Rojas-Corona RR (1989b) Immunohistochemical visualization of afferent nerve terminals in human globus pallidus and ist alteration in neostriatal neurodegenerative disorders. Acta Neuropathol (Berl) 78: 543–550

Hantraye P, Riche D, Maziere M, Isacson O (1990) A primate model of Huntington's disease: behavioural and anatomical studies of unilateral excitotoxic lesions of the caudate-putamen in the baboon. Exp Neurol 108: 91–104

Huang Q, Zhou D, Sapp E, Aizawa H, Ge P, Bird ED, Vonsattel JP, DiFiglia M (1995) Quinolinic acid-induced increases in calbindin D28k immunoreactivity in rat striatal neurons in vivo and in vitro mimic the pattern seen in Huntington's disease. Neuroscience 65: 397–407

Javitch JA, D'Amato RJ, Strittmatter SM, Snyder SH (1985) Parkinsonism inducing neurotoxin, N-methyl-4-phenyl-1,2,3,6-tetrahydropyridine: uptake of the metabolite N-methyl-4-phenylpyridine by dopamine neurons explains selective toxicity. Proc Natl Acad Sci USA 82: 2173–2177

Kato S, Nakamura H (1990) Cytoplasmic argyrophilic inclusions in neurons of pontine nuclei in patients with olivopontocerebellar atrophy: Immunohistochemical and ultrastructural studies. Acta Neuropathol (Berl) 79: 584–594

Lantos PL (1997) Multiple system atrophy. Brain Pathol 7: 1293–1297

Lapchak PL, Araujo DM, Hilt DC, Scheng J, Jiao S (1997) Adenoviral vector-mediated GDNF gene therapy in a rodent lesion model of late stage Parkinson's disease. Brain Res 777: 153–160

Martinez-Serrano A, Björklund A (1996) Protection of the neostriatum against excitotoxic damage by neurotrophin-producing, genetically modified neural stem cells. J Neurosci 16: 4604–4616

Nakao N, Brundi P (1997) Effects of alpha-phenyl-tert-butyl nitrone on neuronal survival and motor function following intrastriatal injections of quinolinate or 3-nitropropionic acid. Neuroscience 76: 749–761

Olanow CW, Kordower JH, Freeman TB (1996) Fetal nigral transplantation as a therapy for Parkinson's disease. Trends Neurosci 19: 102–109

Papp MI, Lantos PL (1992) Accumulation of tubular structures in oligodendroglial and neuronal cells as the basic alteration in multiple system atrophy. J Neurol Sci 107: 172–182

Papp MI, Lantos PL (1994) The distribution of oligodendroglial inclusions in multiple system atrophy and ist relevance to clinical symptomatology. Brain 117: 235–243

Papp MI, Kahn JE, Lantos PL (1989) Glial cytoplasmic inclusions in the CNS of patients with multiple system atrophy (striatonigral degeneration, olivopontocerebellar atrophy and Shy-Drager syndrom). J Neurol Sci 94: 79–100

Perez-Navarro E, Arenas E, Reiriz J, Calvo N, Alberch J (1996) Glial cell line-derived neurotrophic factor protects striatal calbindin-immunoreactive neurons from excitotoxic damage. Neuroscience 75: 345–352

Peschanski M, Cesaro P, Hantraye P (1996) What is needed versus what would be interesting to know bevor undertaking neural transplantation in patients with Huntington's disease. Neuroscience 71: 899–900

Pifl C, Giros B, Caron MG (1993) Dopamine transporter expression confers cytotoxicity to low doses of the parkinsonism-inducing neurotoxin 1-methyl-4-phenylpyridinium. J Neurosci 13: 4246–4253

Probst-Cousin S, Rickert CH, Schmid KW, Gullotta F (1998) Cell death mechanisms in multiple system atrophy. J Neuropathol Exp Neurol 57: 814–821

Quinn NP (1994) Multiple system atrophy. In: Marsden CD, Fahn S (eds) Movement disorders, vol 3. Butterworth-Heinemann, London, pp 262–281

Sauer H, Oertel WH (1994) Progressive degeneration of nigrostriatal dopaminergic neurons following intrastriatal terminal lesions with 6-hydroxydopamine: a combined retrograde tracing and immunocytochemical study in the rat. Neuroscience 59: 401–415

Schulz JB, Hensha DR, Jenkin BG, Ferrant RJ, Kowal NW, Rose BR, Beal M (1994) 3-Acetylpyridine produces age-dependent excitotoxic lesions in the rat striatum. J Cereb Blood Flow Metab 14: 1024–1029

Sonsalla PK, Giovanni A, Sieber BA, Donne KD, Manzino L (1992) Characteristics of dopaminergic neurotoxicity produced by MPTP and metamphetamine. Ann N Y Acad Sci 648: 229–238

Spokes EG, Bannister R, Oppenheimer DR (1979) Multiple system atrophy with autonomic failure — clinical, histological and neurochemical observations on four cases. J Neurol Sci 43: 59–82

Storey E, Hyman BT, Jenkins B, Brouillet E, Miller JM, Rosen BR, Beal MF (1992) 1-Methyl-4-phenylpyridinium produces excitotoxic lesions in rat striatum as a result of impairment of oxidative metabolism. J Neurochem 58: 1975–1978

Sundstrom E, Goldstein M, Jonsson G (1986) Uptake inhibition protects nigro-striatal dopamine neurons from the neurotoxicity of 1-methyl-4-phenylpyridine (MPP+) in mice. Eur J Pharmacol 131: 289–292

Takada M, Kono T (1993) 3-Acetylpyridine neurotoxicity to the nigrostriatal dopamine system in mice. Neurosci Lett 161: 211–214

Tison F, Wenning GK, Daniel SE, Quinn NP (1995) The pathophysiology of parkinsonism in multiple sytem atrophy. Eur J Neurol 2: 435–444

Wenning GK, Ben-Shlomo Y, Magalhaes M, Daniel SE, Quinn NP (1994) Clinical features and natural history of multiple system atrophy. An analysis of 100 cases. Brain 117: 835–845

Wenning GK, Granata R, Laboyrie PM, Quinn NP, Jenner P, Marsden CD (1996a) Reversal of behavioural abnormalities by fetal allografts in a novel rat model of striatonigral degeneration. Mov Disord 11: 522–532

Wenning GK, Laboyrie P, Granata R, Quinn NP, Jenner P, Marsden CD (1996b) Reversal of behavioural abnormalities in a rat model of striatonigral degeneration following intrastriatal transplantation of pure mesencephalic and mesencephalic-striatal cografts. Mov Disord 11: 44 (abstract)

Wenning GK, Tison F, Elliott L, Quinn NP, Daniel SE (1996c) Olivopontocerebellar pathology in multiple system atrophy. Mov Disord 11: 157–162

Wenning GK, Odin P, Morrish P, Rehncrona S, Widner H, Brundin P, Rothwell JC, Brown R, Gustavii B, Hagell P, Jahanshahi M, Sawle G, Björklund A, Brooks DJ,

Marsden CD, Quinn NP, Lindvall O (1997a) Short- and long-term survival and function of unilateral intrastriatal dopaminergic grafts in Parkinson's disease. Ann Neurol 42: 95–107

Wenning GK, Quinn NP (1997b) Parkinsonism. Multiple system atrophy. Baillieres Clin Neurol 6: 187–204

Winkler C, Sauer H, Lee CS, Björklund A (1996) Short-term GDNF treatment provides long-term rescue of lesioned nigral dopaminergic neurons in a rat model of Parkinson's disease. J Neurosci 16: 7206–7215

Wüllner U, Young AB, Penney JB, Beal MF (1994) 3-Nitropropionic acid toxicity in the striatum. J Neurochem 63: 1772–1781

Wüllner U, Weller N, Groscurt P, Loschmann PA, Schulz JB, Müller I, Klockgether T (1997) Evidence for an active type of cell death with ultrastructural features distinct from apoptosis: the effects of 3-acetylpyridine neurotoxicity. Neuroscience 81: 721–734

Yurek DM, Collier TJ, Sladek JR jr (1990) Embryonic mesencephalic and striatal co-grafts: development of grafted dopamine neurons and functional recovery. Exp Neurol 109: 191–199

Authors' address: Dr. G.K. Wenning, Neurological Research Laboratory, Department of Neurology, University Hospital, Anichstrasse 35, A-6020 Innsbruck, Austria

Striatal reconstruction by striatal grafts

S. B. Dunnett

MRC Cambridge Centre for Brain Repair, University of Cambridge, Cambridge,
United Kingdom

Summary. It is now well established that striatal lesions induce motor and cognitive deficits in rats, and that grafts of embryonic striatal tissue can survive, integrate into the lesioned host brain and alleviate the behavioural deficits in both motor and cognitive spheres. How? Since normal striatal function is dependent upon it's integration within a connected cortical-subcortical neuronal circuitry, and the deficits following striatal damage appear to reflect a "disconnexion" syndrome, the observation of recovery suggests that the grafts re-establish a connected circuitry within the host brain. Evidence to corroborate or refute this hypothesis, in comparison with a less-specific mechanism (or mechanisms) of recovery, is considered, including anatomical, electrophysiological and neurochemical demonstrations of functional circuit reconstruction in the host brain by striatal tissue transplants.

Introduction

It is now well established that, under appropriate conditions, grafts of embryonic tissues can survive transplantation into the adult mammalian brain and exert a functional influence on the host animal (Dunnett and Björklund, 1994b). Functional recovery associated with traumatic injury, neurodegenerative damage and disease has been demonstrated following grafts implantation in a wide variety of CNS systems in rodents and primates, and more recently has seen experimental trials of clinical application in Parkinson's disease, Huntington's disease and spinal cord injury (Lindvall, 1997; Olanow et al., 1996; Philpott et al., 1997; Pappas et al., 1997; Pochon et al., 1996).

By contrast to their clear-cut empirical efficacy, the mechanisms by which grafts exert their functional effects have been more widely debated. Thus, for example, in early studies of damage and repair of dopamine and cholinergic systems in the rat, functional recovery was observed to be accompanied by dopaminergic and cholinergic fibre ingrowth from embryonic tissues into the host brain, and it was (perhaps naively) assumed that the functional effects were a consequence of the graft-derived restitution of a regulatory reinnervation of deafferented targets in the host brain (Björklund et al.,

1987). Although this may be the case, it has subsequently been established that grafts might exert their effects via a variety of different mechanisms (Björklund et al., 1987; Dunnett and Björklund, 1994a):

Non-specific effects of transplantation surgery.
Trophic grafts. Graft tissues may release trophic molecules that stimulate intrinsic repair processes in the host brain.
Pharmacological grafts. The graft tissues secrete deficient neurochemicals, enzymes, hormones or transmitters, that exert their effects pharmacologically in the host brain.
Bridge grafts. The graft tissues stimulate sprouting and provide a substrate for axon regeneration to remote targets in the adult brain.
Tonic reinnervation. Grafted neurons reinnervate deafferented targets and restore tonic activation of host circuits, but do not relay temporally or spatially patterned information.
Circuit reconstruction. The grafted tissues establish reciprocal afferent and efferent connections with the host brain, and become integrated within a functional host neuronal circuitry.

Although the ultimate goal of transplantation may be to reconstruct the neuronal circuitry of the brain, it has turned out that most observations of functional recovery in most model systems can be fully explained by one or other of the less specific mechanisms outlined above (Björklund et al., 1987; Dunnett and Björklund, 1994a).

And yet reconstruction remains the ultimate challenge. For example, in spinal cord injury, full recovery is most likely to be achieved if we can restore the full ascending and descending relay of information between spinal cord segments and the sensory and motor centres in the forebrain and brainstem. In spite of a number of optimistic recent developments (Cheng et al., 1996; Li et al., 1997), such reconstruction has not yet been achieved in spinal cord injury. However, clear evidence is now accumulating that such reconstruction is possible under some circumstances; perhaps the clearest evidence is seen in the context of striatal repair with striatal grafts. Evidence in support of this strong hypothesis is reviewed in this chapter.

The neostriatum

The neostriatum is the largest nucleus of the basal ganglia, the high level motor nuclei of the telencephalon. The neostriatum is subdivided into caudate nucleus and putamen in primates and carnivores but is a single undifferentiated nucleus in rodents. The major inputs are from the neocortex, thalamus and a variety of brainstem regulatory systems (including the pars compacta of substantia nigra and the raphé nucleus), and the major outputs are to downstream nuclei of the basal ganglia, including the internal and external segments of the globus pallidus and the pars reticulata of the substantia nigra (see Fig. 1) (Nauta and Domesick, 1984). The predominant cell type of the

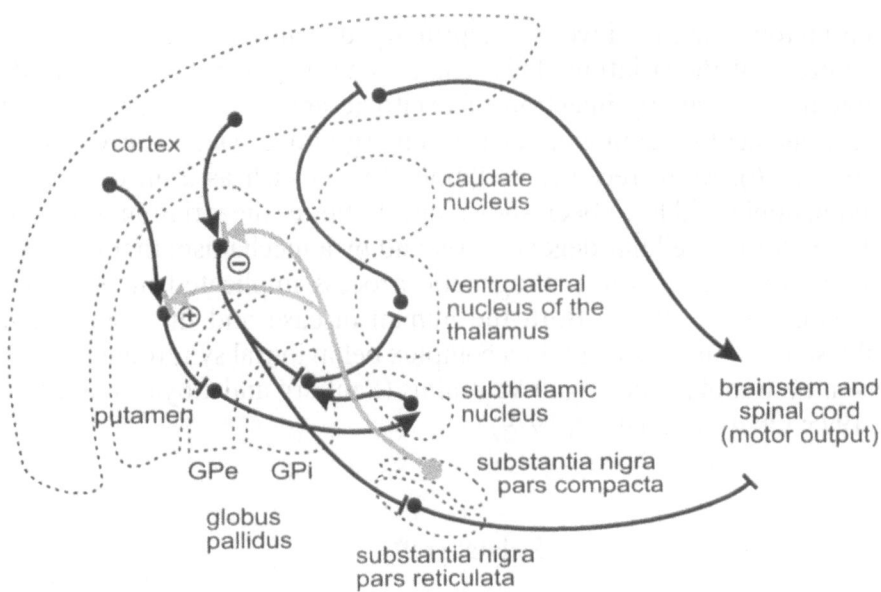

Fig. 1. Circuitry of the basal ganglia. Cortical inputs to the neostriatum (caudate nucleus and putamen) are glutamatergic; the striatal projection neurons to globus pallidus, subthalamic nucleus and substantia nigra pars reticulata are inhibitory and GABAergic. Dopaminergic afferents from the substantia nigra pars compacta have a mixed influence, inhibitory on the outputs of the "direct" pathway to the globus pallidus internal segment and substantia nigra pars reticulata, but excitatory on the outputs of the "indirect" pathway that relays via the external segment of the globus pallidus to subthalamic nucleus and back to the internal segment

neostriatum is the medium spiny neuron. These neurons comprise 90–95% of the total neuronal population of the striatum, they receive all major inputs, and they are the major output projection neurons. The medium spiny neurons are inhibitory, using GABA as their primary transmitter; they co-localise enkephalin in the pallidal projection and dynorphin and substance P in the pars reticulata projection, and they can most easily and distinctively be visualised using immunoreactivity against the dopamine and adenosine monophosphate-related phosphoprotein, DARPP-32 (Ouimet et al., 1984). The remaining neurons are medium and large spiny and aspiny interneurons, using acetylcholine, somatostatin, neuropeptide Y as transmitters and exhibiting NADPH-diaphorase and nitric oxide synthase immunoreactivity (Pasik et al., 1979).

The internal organisation of the striatum has been characterised as having a distinctly heterogeneous organisation in terms of the topography of cortical projections (Alexander and Crutcher, 1990), the striosome/matrix compartmentation in terms of a variety of neurochemical and receptor markers (Graybiel, 1984), and direct vs. indirect projections of the outputs via the pallidal and subthalamic output pathways (DeLong, 1990). This organisation has recently been systematically reviewed elsewhere (Dunnett and Everitt, 1998).

Huntington's disease involves a primary degeneration of the medium spiny neurons of the striatum. This striatal pathology can be reproduced in experimental animals by injection of excitoxic amino acids such as kainic acid, ibotenic acid or quinolinic acid (Sanberg and Coyle, 1984; Coyle and Schwarcz, 1976). More recently metabolic toxins such as 3-nitroproprionic acid and malonic acid have been shown also to target the striatum, producing a similar pattern of cellular degeneration and by a mechanism that may more closely approximate to the pathogenetic process implicated in the human disease (Beal et al., 1993). Both the human disease and the experimental striatal lesions in animals result in a complex behavioural syndrome involving deficits in both motor and cognitive realms (Sanberg and Coyle, 1984; Houk et al., 1995; Öberg and Divac, 1979).

Striatal grafts

The techniques for striatal cell transplantation are relatively straightforward. In the standard cell suspension method (Schmidt et al., 1981), striatal precursors are dissected from the embryonic ganglionic eminence at the appropriate developmental age (E14–15 in rats), incubated in trypsin, washed and dissociated by trituration to form a concentrated suspension of cells that is then stereotaxically injected into the host striatum, most typically homotopically into the site of a striatal lesion (Dunnett and Björklund, 1992). Such grafts have routinely been seen to survive well, and to integrate into the host brain as determined by a combination of anatomical, physiological, neurochemical and behavioural methods (Wictorin, 1992; Björklund et al., 1994).

Functional circuits studied behaviourally

The first behavioural tests of striatal grafts were the demonstration of recovery on simple motor behaviours, such as locomotor hyperactivity (Deckel et al., 1983; Isacson et al., 1986), and indeed striatal grafts are effective in alleviating deficits on a range of simple and more complex motor tests, such as rotation (Dunnett et al., 1988; Lu et al., 1993), skilled paw use (Dunnett et al., 1988; Montoya et al., 1990) and lateralised reaction time tasks in operant chambers (Döbrössy and Dunnett, 1999; Mayer et al., 1992).

Of greater interest is the fact that these grafts can also alleviate performance on a number of cognitive tasks sensitive to the integrity of prefrontal cortical systems. This was first demonstrated on the classic delayed alternation task in the T maze (Isacson et al., 1986), in which study it was shown that striatal grafts must be placed into the striatum to be effective, whereas implants in a pallidal target placement were without significant effect. Subsequent studies have demonstrated the capacity of striatal grafts to alleviate deficits in other T maze alternation tasks (Deckel et al., 1986), passive avoidance learning tests (Piña et al., 1994; Koide et al., 1993) and differential reinforcement of low rate responding in operant tasks (Reading et al., 1995). In several of these studies the recovery was seen to be specifically dependent upon the implantation of striatal tissues and was not reproduced by grafts of other tissue types (Lu et al., 1993; Piña et al., 1994; Montoya et al., 1990).

An important feature of the connections of the striatum is that it comprises an integral component of the output pathways of the neocortex, particularly the areas of association cortex involved in mediating higher, cognitive functions (see Fig. 1). This was first described in terms of the neostriatum as a key component in the frontal projection system in primates (Rosvold, 1972). Thus, lesions in discrete areas of prefrontal cortex produce distinctive patterns of impairment on cognitive tests of spatial and object discrimination, reversal, short-term memory and planning, which are reproduced by lesions in the related areas of the striatum to which each cortical area projects (Divac et al., 1967). Critically, deficits associated with lesions in the prefrontal cortex itself are reproduced by lesions in mediodorsal thalamic nuclei that project to prefrontal areas as well as by lesions in the major cortical output targets in the caudate nucleus and globus pallidus.

Such a pattern of results is described as a "disconnexion syndrome" and is the defining feature of a set of nuclei organised in a network of serial projections (see Fig. 2) (Geschwind, 1965b; Geschwind, 1965a). Critically, damage in such a network cannot be repaired simply by restoring activation in the output or target nuclei (as works for example by dopaminergic activation of deafferented terminals in Parkinson's disease), but only by restoring the relay of information flow through the serial structures of the network. It is informative that whereas l-dopa (as well as dopaminergic agonists and nigral grafts) has proved highly effective in alleviating the deficits of Parkinson's disease, no similar pharmacological strategy has ever been found effective for the cognitive deficits associated with prefrontal damage, striatal damage or Huntington's disease itself.

The most plausible hypothesis to account for the observed recovery in grafted animals on the "prefrontal" cognitive tests is therefore to suggest that the grafts are actually restoring the relay of functional connections within cortical-striatal-pallidal circuits of the forebrain (see Fig. 2). Such a strong hypothesis requires more direct demonstration than theoretical parsimony alone.

Striatal graft morphology

When first examined using acetylcholinesterase as a simple stain for striatal neuropil, striatal grafts were seen to exhibit a distinctive patchy appearance. It turns out that this is attributable to the aggregation within grafts of striatal-like cells in patches (the "P zones') interspersed with a matrix of neurons that

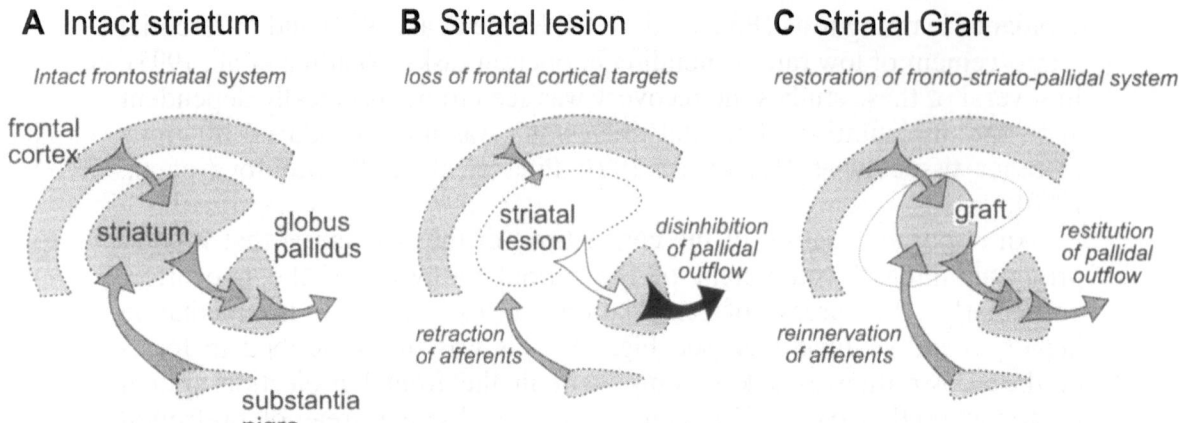

A Intact striatum   **B** Striatal lesion   **C** Striatal Graft

Fig. 2. "Disconnexion syndromes" in the basal ganglia. **A** Simplified circuitry emphasising nigral regulation of a cortico-striatal-pallidal relay of organised plans of action. **B** Striatal lesions disrupt relay of cortical information to downstream motor targets. **C** Striatal grafts not only restore striatal inhibition of pallidal outputs but also restore relay of nigral and cortical information to those same targets

are essentially non-striatal in their morphology (the NP zones'). This pattern was first established on the basis of immunohistochemical staining of neurotransmitter types (Graybiel et al., 1989) and has subsequently been confirmed using other markers based on in situ hybridisation, anterograde and retrograde tracing, Golgi staining and receptor binding (Campbell et al., 1992; Wictorin et al., 1989a; Clarke et al., 1994; Graybiel et al., 1990; Liu et al., 1990; Sirinathsinghji et al., 1993b). Thus, although healthy neurons are seen throughout the grafts in Nissl stained preparations, striatal medium spiny neurons with characteristic DARPP-32 staining patterns as well as large cholinergic and NADPH-diaphorase positive interneurons cluster together in patches, interspersed with NP zones which are comprised of cells with the Golgi and staining characteristics of neocortical, pallidal and other forebrain nuclei. It remains unclear how this pattern of self-aggregation of like cells within striatal grafts arises when they had been implanted as a dissociated and fully intermixed suspension. Two possible mechanisms relate to a local migration of neurons under cell surface-mediated, attractive and repulsive influences, and to a differential survival of cells associated with critical levels of selective trophic factors.

Nevertheless, for present purposes, the critical observation is that neurons of all main striatal subpopulations survive in striatal grafts (Clarke et al., 1994; Helm et al., 1990), and re-establish a striatal like internal organisation within the host brain. Moreover, although data is still sparse, there is growing evidence that functional recovery in striatally grafted animals relates to the survival and size of the P zone compartment rather than to total graft volume (Fricker et al., 1997; Nakao et al., 1996), suggesting that the recovery may reflect a specific process of striatal repair.

Striatal graft connectivity

To be effective a graft must not only survive and replace the missing populations of striatal neurons, it must integrate with the host brain. In particular, if we are to see recovery from a disconnexion syndrome, then one critical question is the extent to which cortico-striatal and striato-pallidal connections are actually restored between the grafts and the host brain.

There is now extensive evidence, in particular in the work of Klas Wictorin, that striatal grafts routinely integrate well. He used a combination of anterograde and retrograde tracing methods to establish the reformation of afferents from the host cortex, thalamus, substantia nigra and raphe nucleus, and rich efferent outgrowth routinely to the globus pallidus and in many cases to the substantia nigra pars reticulata (see Fig. 3) (Wictorin, 1992). Thus, all main inputs and outputs of the normal striatum appear to be reformed within striatal grafts at the global level. Moreover, retrograde tracing indicates that the reconnectivity follows the patchwork internal organisation of grafts: the pallidal and nigral efferents appear to originate from the neurons of the P zones, and at least the nigral afferents in tyrosine hydroxylase immunohistochemistry selectively reinnervate the striatal P zone compartment (Wictorin et al., 1989b).

Within the P zones, connections are re-established at morphologically appropriate synaptic junctions (Wictorin et al., 1989b). In a detailed study, Clarke and Dunnett (Clarke and Dunnett, 1993) combined retrograde tracing, anterograde degeneration, Golgi impregnation and neurotransmitter immunohistochemistry to identify specific morphological components of the circuitry within the P zones of striatal grafts at the ultrastructural level. We were able to demonstrate most pairwise and three-way combinations

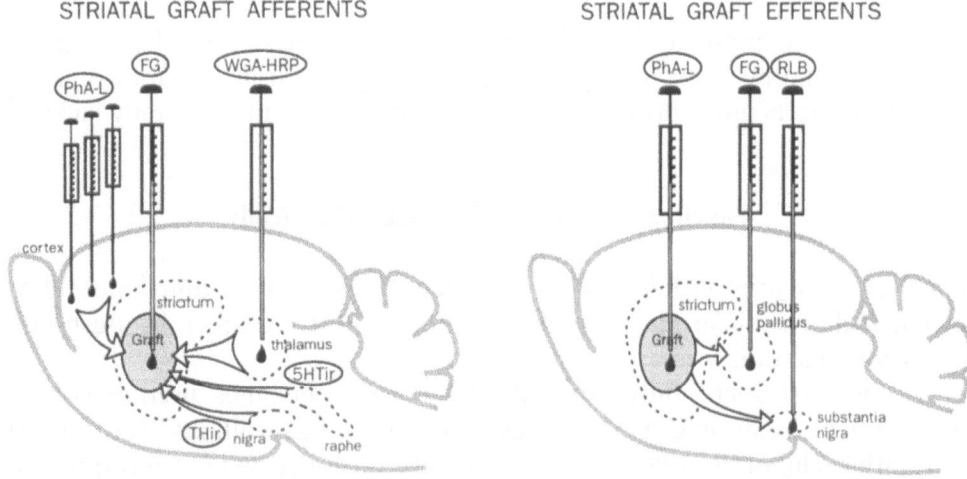

Fig. 3. Connectivity of striatal grafts. Anterograde and retrograde axon tracing experiments in combination with immunohistochemistry, indicate reformation of the major afferent and efferent pathways of the normal striatum into and out of striatal grafts. Redrawn from Wictorin (1992)

between: (a) TH-immunoreactive dopamine terminals making appropriate symmetric synapses onto the necks of spines and shafts of dendrites; (b) degenerating corticostriatal terminals making appropriate asymmetric terminals onto the heads of spines, of neurons that were (c) of medium spiny morphology based on Golgi staining, (d) were GABAergic based on GAD-immunoreactivity, and (e) that were output neurons projecting to the globus pallidus based on retrograde transport of horseradish peroxidase.

Although electron microscopy cannot yield fully quantitative data, these results clearly demonstrate reconstruction of an anatomical circuitry of the type that would be required to restore a cortico-striato-pallidal relay of functional information within a prefrontal system for the organisation of complex task behaviours.

Functional circuits studied physiologically

The functional relay of neuronal information can most readily be observed electrophysiologically. Rutherford et al first demonstrated monosynaptic responses in grafted striatal neurons following stimulation of the afferents from the neocortex in tissue slices cut at an oblique angle to incorporate the whole cortical striatal projection within the slice (Rutherford et al., 1987).

Subsequent studies have provided elaborated these observations in vivo, and confirmed short latency excitatory responses from functional cortical and thalamic inputs (Wilson et al., 1990; Walsh et al., 1988). Nevertheless, the responses were not fully normal; although the short latency response reflecting the direct monosynaptic input was similar to that seen in the normal striatum, the subsequent longer latency hyperpolarisation and rebound excitation was markedly reduced in grafted neurons, suggesting a deficit in polysynaptic connections that would be required for full collateral (interneuronal) regulation (Xu et al., 1991).

Apart from these few studies, detailed electrophysiological analysis of striatal graft inputs is sparse, and the further analysis of trans-synaptic integration of the grafts into a host-graft-host circuitry as yet remains totally lacking.

Functional circuits studied neurochemically

An alternative approach is to use neurochemical markers of functional relay, in particular using in vivo sampling of transmitter turnover by push-pull perfusion, microdialysis or electrochemistry.

In the first such study, Sirinathsinghji and colleagues used push-pull perfusion to monitor GABA turnover in the globus pallidus (see Fig. 4) (Sirinathsinghji et al., 1988). Normal levels of turnover were reduced to 5% of baseline level by a striatal lesion, and restored to approximately 34% of baseline in rats with striatal grafts, presumably derived from terminals of grafted GABAergic neurons, the axons of which had regrown into the deafferented pallidus. Of even greater interest in this study was the demon-

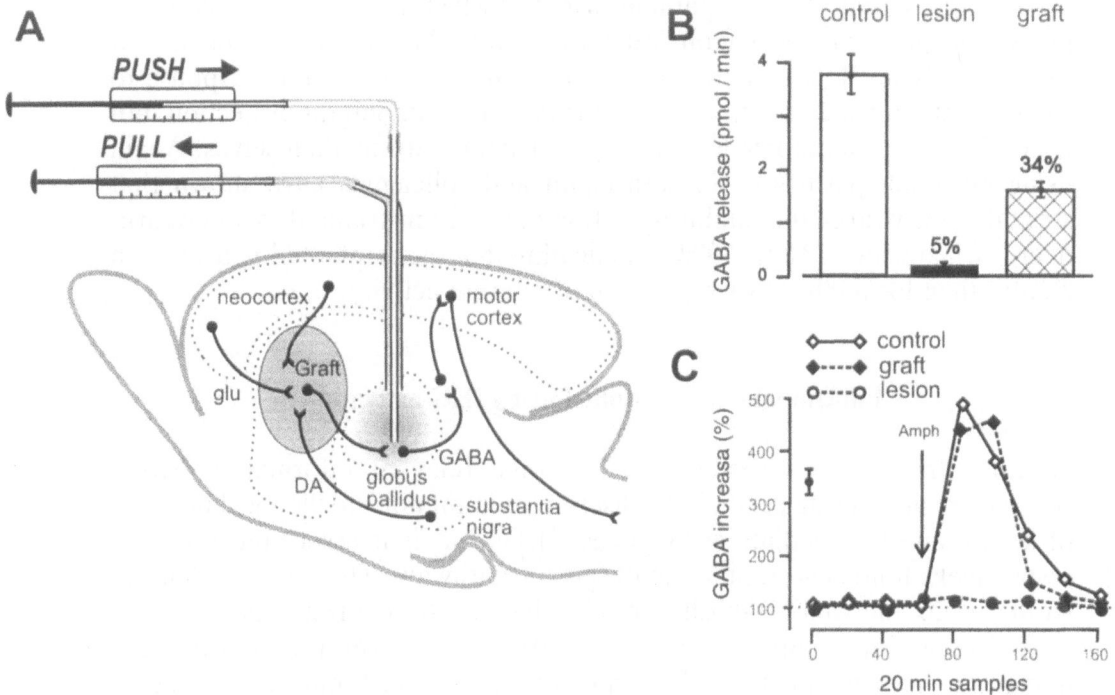

Fig. 4. A Schematic illustration of push-pull perfusion analysis of GABA turnover in the pallidum of normal, striatal lesion and striatal graft rats. **B** Striatal lesions almost completely abolish baseline levels of GABA turnover in the globus pallidus, which is partially restored by striatal grafts. **C** Dopamine activation of the striatum by injection of amphetamine (arrow) induces a 5 fold surge of GABA release in the pallidum, lasting approx. 40 mins. This response is completely abolished in the lesion animals. The response is completely restored in the grafted animals, in terms both of its magnitude (as a proportional measure against the lower baseline) and of its duration

stration that peripheral injection of amphetamine induces a marked surge of pallidal GABA release in the intact animals, due to drug induced release of dopamine from nigrostriatal terminals activating the striatopallidal GABAergic projection neurons. This response was completely abolished in rats with striatal lesions but fully restored in rats with striatal grafts (Sirinathsinghji et al., 1988). This study provided the first clear evidence that information relayed via an afferent input to a striatal graft can be transduced by graft neurons and relayed back to targets within the host brain, which is the essential criterion of circuit reconstruction. The ability of striatal grafts to regulate GABA turnover on the basis of their afferents has subsequently been replicated by Campbell and colleagues (Campbell et al., 1993).

A similar technique has been used to investigate local interneuronal regulation of transmitter turnover. Thus, CCK infused into the striatum inhibits dopamine turnover from nigrostriatal terminals. This is believed to be reflect the fact that intrinsic CCK afferents from the claustrum, amygdala, frontal and parietal cortex synapse on CCK receptors on striatal medium spiny neurons (Myer et al., 1981; Beresford et al., 1987), which provide a feedforward

GABAergic inhibition of dopamine turnover which is believed to act at the presynaptic level via local axon collaterals within the striatum rather than at the cell body level via striatonigral projects since the dopamine response to CCK is not abolished by knife cut lesions of the striatonigral projection, but is abolished by excitotoxic lesions of the striatal neurons themselves. Again using push-pull perfusion, Sirinathsinghji and colleagues have shown that striatal grafts restore the inhibitory action of CCK on striatal dopamine turnover (Sirinathsinghji et al., 1993a), indicating that the implanted neurons can restore their local inhibitory regulation of afferent activity.

Functional circuits studied using IEG activation

A third strategy for investigating functional relay of information within striatal circuits is using in situ hybridisation or immunohistochemical markers of activation of immediate early genes (IEGs) or their associated proteins, respectively, following stimulation of specific pathways. This was first demonstrated using the well established model that striatal neurons show massive induction of c-fos, c-jun, krox and other IEGs over several hours following dopaminergic activation using the stimulant drugs amphetamine or cocaine. A similar activation of IEG activity is seen in striatal grafts after administration of both stimulants (Dragunow et al., 1991; Liu et al., 1991; Mandel et al., 1992) and dopamine antagonists (Dragunow et al., 1990), reflecting the fact that IEG induction reflects major changes (rather than just increases) in neuronal activity, and indicating that the change in the activity of afferent host neurons is indeed transduced by graft neurons. Moreover, Fos induction is observed relatively selectively in neurons of the P-zones, being the neurons within the grafts that both exhibit striatal-like phenotype and receive a rich innervation from host dopamine terminals (Liu et al., 1991).

Fos immunoreactivity has also been employed to demonstrate functional transduction of inputs from host neocortex. Thus, Labandeira-Garcia and colleagues have shown a massive induction of c-fos activity in striatal grafts following electrical stimulation of the frontal cortex. Moreover Fos-positive neurons were strongly colocalised in the P-zones of the grafts as determined by DARPP-32 staining in adjacent sections (Labandeira-Garcia and Guerra, 1994; Labandeira-Garcia et al., 1994). This is important, since whereas an early study of cortical fibre ingrowth suggested that these projections may be rather diffuse and not very extensive (Wictorin and Björklund, 1989), the IEG activation studies indicate that even if the density of terminals is only partial the activity relayed via that projection is received and effectively transduced by the relevant target neurons.

Summary

The striatal graft model has provided a powerful paradigm for evaluating the capacity of neural grafts not only to replace deficient neuroactive chemicals (be they neurotransmitters, neurohormones, growth factors or other neuroactive molecules), functioning as sophisticated pharmacological minipumps, but also to reconstruct damaged neuronal pathways in the brain and restore functions that are dependent upon information transfer within connected neural networks. Ultimately, it is logically impossible to prove the hypothesis that striatal grafts function by reconstruction of the neuronal pathways of the brain (Popper, 1963), but the hypothesis is corroborated by anatomical, electrophysiological, pharmacological, and neurochemical data that the necessary connections are reformed, and it remains the most plausible explanation for recovery of behavioural deficits known to be dependent on the integrity of a prefrontal system involving the relay in patterned information through a cortical-striatal-pallidal-thalamic-cortical open loop network of connections.

Acknowledgements

I acknowledge the contributions of many scientific friends, colleagues and students who have contributed to this work, including Anders Björklund, Susan Iversen, Rusty Gage, Ole Isacson, Debby Clarke, Dalip Sirinathsinghji, Ann Graybiel, Fu-Chin Liu, Chris Montoya, Paul Reading, Nadia Haque, Rose Fricker and Ed Torres. My own research reported in this review has been supported by the Medical Research Council, the Wellcome Trust, the Hereditary Disease Foundation and NATO.

References

Alexander GE, Crutcher MD (1990) Functional architecture of basal ganglia circuits: neural substrates of parallel processing. Trends Neurosci 13: 266–271

Beal MF, Hyman BT, Koroshetz W (1993) Do defects in mitochondrial energy metabolism underlie the pathology of neurodegenerative diseases? Trends Neurosci 16: 125–131

Beresford IJM, Hall MD, Clark CR, Hill RG, Hughes J, Sirinathsinghji DJS (1987) Striatal lesions and transplants demonstrate that cholecystokinin receptors are localized on intrinsic striatal neurones: a quantitative autoradiographic study. Neuropeptides 10: 109–136

Björklund A, Lindvall O, Isacson O, Brundin P, Wictorin K, Strecker RE, Clarke DJ, Dunnett SB (1987) Mechanisms of action of intracerebral neural implants — studies on nigral and striatal grafts to the lesioned striatum. Trends Neurosci 10: 509–516

Björklund A, Campbell K, Sirinathsinghji DJS, et al (1994) Functional capacity of striatal transplants in the rat Huntington model. In: Dunnett SB, Björklund A (eds) Functional neural transplantation. Raven Press, New York, pp 157–195

Campbell K, Wictorin K, Björklund A (1992) Differential regulation of neuropeptide mRNA expression in intrastriatal striatal transplants by host dopaminergic afferents. Proc Natl Acad Sci USA 89: 10489–10493

Campbell K, Kalén P, Wictorin K, Lundberg C, Mandel RJ, Björklund A (1993) Characterization of GABA release from intrastriatal striatal transplants: dependence on host-derived afferents. Neuroscience 53: 403–415

Cheng H, Cao YH, Olson L (1996) Spinal cord repair in adult paraplegic rats — partial restoration of hind-limb function. Science 273: 510–513

Clarke DJ, Dunnett SB (1993) Synaptic relationships between cortical and dopaminergic inputs and intrinsic GABAergic systems within intrastriatal striatal grafts. J Chem Neuroanat 6: 147–158

Clarke DJ, Wictorin K, Dunnett SB, et al (1994) Internal composition of striatal grafts: light and electron microscopy. In: Percheron G, McKenzie JS, Féger J (eds) The basal ganglia IV, New ideas on structure and function. Plenum Press, New York, pp 189–196

Coyle JT, Schwarcz R (1976) Lesions of striatal neurones with kainic acid provides a model for Huntington's chorea. Nature 263: 244–246

Deckel AW, Robinson RG, Coyle JT, Sanberg PR (1983) Reversal of long-term locomotor abnormalities in the kainic acid model of Huntington's disease by day 18 fetal striatal implants. Eur J Pharmacol 92: 287–288

Deckel AW, Moran TH, Coyle JT, Sanberg PR, Robinson RG (1986) Anatomical predictors of behavioral recovery following fetal striatal transplants. Brain Res 365: 249–258

DeLong MR (1990) Primate models of movement disorders of basal ganglia origin. Trends Neurosci 13: 281–285

Divac I, Rosvold HE, Szwarcbart MK (1967) Behavioral effects of selective ablation of the caudate nucleus. J Comp Physiol Psychol 63: 184–190

Döbrössy MD, Dunnett SB (1999) Striatal grafts alleviate deficits in response execution in a lateralised reaction time task. Brain Res Bull (in press)

Dragunow M, Williams M, Faull RLM (1990) Haloperidol induces Fos and related molecules in intrastriatal grafts derived from fetal striatal primordium. Brain Res 530: 309–311

Dragunow M, Faull RLM, Waldvogel HJ, Williams MN, Leah J (1991) Elevated expression of jun and fos-related proteins in transplanted striatal neurons. Brain Res 558: 321–324

Dunnett SB, Björklund A (1992) Neural transplantation: a practical approach. IRL Press, Oxford

Dunnett SB, Björklund A (1994a) Mechanisms of function of neural grafts in the injured brain. In: Dunnett SB, Björklund A (eds) Functional neural transplantation. Raven Press, New York, pp 531–567

Dunnett SB, Björklund A (1994b) Functional neural transplantation. Raven Press, New York

Dunnett SB, Everitt BJ (1998) Topographic factors affecting the functional viability of dopamine-rich grafts in the neostriatum. In: Freeman TB, Kordower JH (eds) Cell transplantation for neurological disorders. Humana Press, Totowa, NJ, pp 135–169

Dunnett SB, Isacson O, Sirinathsinghji DJS, Clarke DJ, Björklund A (1988) Striatal grafts in rats with unilateral neostriatal lesions. III. Recovery from dopamine-dependent motor asymmetry and deficits in skilled paw reaching. Neuroscience 24: 813–820

Fricker RA, Torres EM, Hume SP, Myers R, Opacka-Juffry J, Ashworth S, Dunnett SB (1997) The effects of donor stage on the survival and function of embryonic striatal grafts. II. Correlation between positron emission tomography and reaching behaviour. Neuroscience 79: 711–722

Geschwind N (1965a) Disconnexion syndromes in animals and man. Part I. Brain 88: 237–294

Geschwind N (1965b) Disconnexion syndromes in animals and man. Part II. Brain 88: 585–644

Graybiel AM (1984) Neurochemically specified subsystems in the basal ganglia. In: Ciba Foundation Symposium 107 (ed) Functions of the basal ganglia. Pitman, London, pp 114–149

Graybiel AM, Liu FC, Dunnett SB (1989) Intrastriatal grafts derived from fetal striatal primordia. 1. Phenotypy and modular organization. J Neurosci 9: 3250–3271

Graybiel AM, Liu FC, Dunnett SB (1990) Cellular reaggregation in vivo: modular patterns in intrastriatal grafts derived from fetal striatal primordia. Prog Brain Res 82: 401–405

Helm GA, Palmer PE, Bennett JP (1990) Fetal neostriatal transplants in the rat: a light and electron microscopic golgi study. Neuroscience 37: 735–756

Houk JC, Davis JL, Beiser DG (1995) Models of information processing in the basal ganglia. MIT Press, Cambridge MA

Isacson O, Dunnett SB, Björklund A (1986) Graft-induced behavioral recovery in an animal model of Huntington disease. Proc Natl Acad Sci USA 83: 2728–2732

Koide K, Hashitani T, Aihara N, Mabe H, Nishino H (1993) Improvement of passive avoidance task after grafting of fetal striatal cell suspensions in ischemic striatum in the rat. Rest Neurol Neurosci 5: 205–214

Labandeira-Garcia JL, Guerra MJ (1994) Cortical stimulation induces fos expression in intrastriatal striatal grafts. Brain Res 652: 87–97

Labandeira-Garcia JL, Tobio JP, Guerra MJ (1994) Comparison between normal developing striatum and developing striatal grafts using drug-induced Fos expression and neuron-specific enolase immunohistochemistry. Neuroscience 60: 399–415

Li Y, Field PM, Raisman G (1997) Repair of adult rat corticospinal tract by transplants of olfactory ensheathing cells. Science 277: 2000–2002

Lindvall O (1997) Neural transplantation: a hope for patients with Parkinson's disease? Neuroreport 8(14): iii–x

Liu FC, Graybiel AM, Dunnett SB, Baughman RW (1990) Intrastriatal grafts derived from fetal striatal primordia. 2. Reconstitution of cholinergic and dopaminergic systems. J Comp Neurol 295: 1–14

Liu FC, Dunnett SB, Robertson HA, Graybiel AM (1991) Intrastriatal grafts derived from fetal striatal primordia. 3. Induction of modular patterns of fos-like immunoreactivity by cocaine. Exp Brain Res 85: 501–506

Lu SY, Pixley SK, Emerich DF, Lehman MN, Norman AB (1993) Effect of fetal striatal and astrocyte transplants into unilateral excitotoxin-lesioned striatum. J Neur Transplant Plast 4: 279–287

Mandel RJ, Wictorin K, Cenci MA, Björklund A (1992) Fos expression in intrastriatal striatal grafts: regulation by host dopaminergic afferents. Brain Res 583: 207–215

Mayer E, Brown VJ, Dunnett SB, Robbins TW (1992) Striatal graft-associated recovery of a lesion-induced performance deficit in the rat requires learning to use the transplant. Eur J Neurosci 4: 119–126

Montoya CP, Astell S, Dunnett SB (1990) Effects of nigral and striatal grafts on skilled forelimb use in the rat. Prog Brain Res 82: 459–466

Myer DK, Beinfeld MC, Oertel WH, Brownstein MJ (1981) Origin of the cholecystokinin-containing fibers in the rat caudatoputamen. Science 215: 187–188

Nakao N, Grasbon-Frodl EM, Widner H, Brundin P (1996) DARPP-32-rich zones in grafts of lateral ganglionic eminence govern the extent of functional recovery in skilled paw reaching in an animal model of Huntington's disease. Neuroscience 74: 959–970

Nauta WJH, Domesick VB (1984) Afferent and efferent relationships of the basal ganglia. In: Ciba Foundation Symposium 107 (ed) Functions of the basal ganglia. Pitman, London, pp 3–23

Olanow CW, Kordower JH, Freeman TB (1996) Fetal nigral transplantation as a therapy for Parkinson's disease. Trends Neurosci 19: 102–109

Ouimet CC, Miller PE, Hemmings HC, Walaas SI, Greengard P (1984) DARPP-32, a dopamine- and adenosine-3':5'-monophosphate-regulated phosphoprotein enriched in dopamine-innervated brain regions. J Neurosci 4: 111–124

Öberg RGE, Divac I (1979) Cognitive functions of the neostriatum. In: Divac I, Öberg RGE (eds) The neostriatum. Pergamon Press, Oxford

Pappas GD, Lazorthes Y, Bès JC, Tafani M, Winnie AP (1997) Relief of intractable cancer pain by human chromaffin cell transplants: Experience at two medical centers. Neurol Res 19: 71–77

Pasik P, Pasik T, DiFiglia M (1979) The internal organization of the neostriatum in mammals. In: Divac I, Öberg RGE (eds) The neostriatum. Pergamon Press, Oxford, pp 5–36

Philpott LM, Kopyov OV, Lee AJ, Jacques S, Duma CM, Caine S, Yang M, Eagle KS (1997) Neuropsychological functioning following fetal striatal transplantation in Huntington's chorea: Three case presentations. Cell Transplant 6: 203–212

Piña AL, Ormsby CE, Bermúdez-Rattoni F (1994) Differential recovery of inhibitory avoidance learning by striatal, cortical, and mesencephalic fetal grafts. Behav Neur Biol 61: 196–201

Pochon NAM, Heyd B, Deglon N, Joseph JM, Zurn AD, Baetge EE, Hammang JP, Goddard M, Lysaght M, Kaplan FA, Kato AC, Schluep M, Hirt L, Regli F, Porchet F, De Tribolet N, Aebischer P (1996) Gene therapy for amyotrophic lateral sclerosis (ALS) using a polymer encapsulated xenogenic cell line engineered to secrete hCNTF. Hum Gene Ther 7: 851–860

Popper K (1963) Conjectures and refutations: the growth of scientific knowledge. Routledge, London

Reading PJ, Torres EM, Dunnett SB (1995) Embryonic striatal grafts ameliorate the disinhibitory effects of ventral striatal lesions. Exp Brain Res 105: 76–86

Rosvold HE (1972) The frontal lobe system: cortical-subcortical interrelationships. Acta Neurobiol Exp 32: 439–460

Rutherford A, Garcia-Munoz M, Dunnett SB, Arbuthnott GW (1987) Electrophysiological demonstration of host cortical inputs to striatal grafts. Neurosci Lett 83: 275–281

Sanberg PR, Coyle JT (1984) Scientific approaches to Huntington's disease. CRC Crit Rev Clin Neurobiol 1: 1–44

Schmidt RH, Björklund A, Stenevi U (1981) Intracerebral grafting of dissociated cell suspensions: a new approach for neuronal transplantation to deep brain sites. Brain Res 218: 347–356

Sirinathsinghji DJS, Dunnett SB, Isacson O, Clarke DJ, Kendrick K, Björklund A (1988) Striatal grafts in rats with unilateral neostriatal lesions. II. In vivo monitoring of GABA release in globus pallidus and substantia nigra. Neuroscience 24: 803–811

Sirinathsinghji DJS, Heavens RP, Torres EM, Dunnett SB (1993a) Cholecystokinin-dependent regulation of host dopamine inputs to striatal grafts. Neuroscience 53: 651–663

Sirinathsinghji DJS, Mayer E, Fernandez JM, Dunnett SB (1993b) The localisation of CCK mRNA in embryonic striatal tissue grafts: further evidence for the presence of non-striatal cells. Neuroreport 4: 659–662

Walsh JP, Zhou FC, Hull CD, Fisher RS, Levine MS, Buchwald NA (1988) Physiological and morphological characterization of striatal neurons transplanted into the striatum of adult rats. Synapse 2: 37–44

Wictorin K (1992) Anatomy and connectivity of intrastriatal striatal transplants. Prog Neurobiol 38: 611–639

Wictorin K, Björklund A (1989) Connectivity of striatal grafts implanted into the ibotenic acid-lesioned striatum. 2. Cortical afferents. Neuroscience 30: 297–311

Wictorin K, Clarke DJ, Bolam JP, Björklund A (1989a) Host corticostriatal fibres establish synaptic connections with grafted striatal neurons in the ibotenic acid lesioned striatum. Eur J Neurosci 1: 189–195

Wictorin K, Ouimet CC, Björklund A (1989b) Intrinsic organization and connectivity of intrastriatal striatal transplants in rats as revealed by DARPP-32 immunohistochemistry: specificity of connections with the lesioned host brain. Eur J Neurosci 1: 690–701

Wictorin K, Simerly RB, Isacson O, Swanson LW, Björklund A (1989c) Connectivity of striatal grafts implanted into the ibotenic acid- lesioned striatum. 3. Efferent projecting graft neurons and their relation to host afferents within the grafts. Neuroscience 30: 313–330

Wilson CJ, Xu ZC, Emson PC, Feler C (1990) Anatomical and physiological properties of the cortical and thalamic innervations of neostriatal tissue grafts. Prog Brain Res 82: 417–426

Xu ZC, Wilson CJ, Emson PC (1991) Synaptic potentials evoked in spiny neurons in rat neostriatal grafts by cortical and thalamic stimulation. J Neurophysiol 65: 477–493

Author's address: Stephen Dunnett, MRC Cambridge Centre for Brain Repair, University Forvie Site, Robinson Way, Cambridge CB2 2PY, U.K.

Somatic gene therapy in animal models of Parkinson's disease

M. Bauer[1], **M. Ueffing**[2,3], **T. Meitinger**[2], and **T. Gasser**[1]

[1] Department of Neurology, Klinikum Grosshadern, Ludwig-Maximilians-University,
Munich, Federal Republic of Germany
[2] Department of Medical Genetics, Ludwig-Maximilians-University, Munich,
Federal Republic of Germany
[3] Institute for Molecular Biology and Tumour Genetics, GSF, Munich,
Federal Republic of Germany

Summary. Gene therapy in Parkinson's disease (PD) emerged about 10 years ago but until now, no clinical trials are under way, because most approaches have failed to show long-term therapeutic effects in PD animal models and because safety concerns precluded the use in humans so far. This review tries to give an overview on the development of different strategies in gene therapy in PD animal models and point out new and possibly more successful directions, including the transplantation of neural precursor cells and pig tissue.

Introduction

Parkinson's disease (PD) is a chronic disorder characterised by a progressive degeneration of dopaminergic neurones of the substantia nigra (SN) pars compacta leading to a reduction of striatal dopamine (Forno, 1996). The characteristic motor symptoms including bradykinesia, rigidity and tremor appear when more than 75% of the striatal dopamine has been lost (Duvoisin, 1992; Lange, 1992). The aetiology of PD is still unknown in the vast majority of cases. Symptomatic treatment is available, but so far degeneration of dopaminergic neurones can not be halted and probably not even be slowed significantly.

During the past few years, progress in biomedical sciences made novel gene therapeutic approaches possible. Genes encoding neurotransmitter synthesising enzymes and neurotrophic factors which counteract neurodegenerative processes have been isolated. They provide the basis for recently emerging developments aimed to provide gene therapy for patients with Parkinson's disease.

In principal two approaches for gene transfer into the brain have been established: (1) Genetic manipulation of cultured cells in vitro and subsequent transplantation into the brain (*ex vivo* gene therapy) and (2) direct transfer of genetic material to cerebral target cells *in vivo* (*in vivo* gene therapy).

Both strategies have been applied in order to achieve two fundamentally different goals: Transfer of genes encoding tyrosine hydroxylase (TH), the rate-limiting enzyme in catecholamine biosynthesis, in order to restore the intrastriatal production of dopamine (DA), or alternatively, transfer of genes coding for neurotrophic factors that might protect nigral neurones from degeneration.

The aim of this review is to give an overview of current gene therapeutic strategies in animal models of PD and to critically evaluate the different experimental set-ups.

Ex vivo gene therapy for dopamine replacement

A variety of immortalised cell lines (Horellou et al., 1989, 1990a, 1990b, 1990c; Wolff et al., 1989) and primary cells (Bencsics et al., 1996; Fisher et al., 1991; Jiao et al., 1992a, 1992b; Lundberg et al., 1996; Yoshimoto et al., 1995) cultured and expanded in vitro after isolation from a donor animal have been generated for application as grafts in animal models of Parkinson's disease, and techniques have been established to introduce genetic material into these cells, including calcium phosphate precipitation, electroporation, lipofection and viral transduction.

Cell lines

In early experiments, Hefti et al. (1985) transplanted PC12 cells, derived from a rat pheochromocytoma cell line into the striatum of unilaterally 6-hydroxydopamine (6-OHDA) lesioned rats. They reported an attenuation of apomorphine-induced circling behaviour (Hefti et al., 1985). Several years later, first experiments with genetically modified cell lines have been published: 208F fibroblasts were transduced with a retrovirus vector containing the cDNA for TH. The genetically modified fibroblasts were transplanted into the brain of unilaterally 6-OHDA lesioned rats. Apomorphin-induced rotational behaviour improved for two weeks after grafting (Wolff et al., 1989). In subsequent experiments with other cell lines, including NS20 Y, AtT-20 and RIN, TH transgene expression and reduction of apomorphin-induced rotation asymmetry was observed for the same time period (Horellou et al., 1990a, 1990b).

In retrospect, it is doubtful whether the functional improvement was due to dopamine production of the graft or whether it was due to an indirect effect of the transplantation itself. It has since been shown that reactive astrocytes are stimulated by IL-1 to secret neurotrophic factors which in turn could induce sprouting of dopaminergic neurones in the host striatum (Wang et al., 1994). The transplantation procedure may thus serve as a trauma to brain tissue, resulting in an increased secretion of IL-1 and other factors by microglia, macrophages and lymphocytes (Giulian and Lachman, 1985).

Nevertheless, it soon became clear that the use of immortalised cell lines is limited because they do not survive longer than a few weeks after implantation into the brain of non-immuno-suppressed animals (Horellou et al., 1990a; Wolff et al., 1989a) and because of the tumour forming potential of these cell lines, particularly in immune-suppressed animals (Horellou et al., 1990a, 1990c).

Primary cells

In contrast to immortalised cell lines, primary cells do not form tumours and they have the potential to survive well after intracerebral transplantation for up to several months (Fisher et al., 1991; Jiao et al., 1992a, 1992b). Several cell types have been used for *ex vivo* gene therapy in animal models of PD, including primary fibroblasts (Bencsics et al., 1996, Kawaja and Gage, 1992), myoblasts (Jiao and Wolff, 1992b; Partridge and Davies, 1995), glial cells (Lundberg et al., 1996) and mesencephalic cultures (Zawada et al., 1998; Zhu et al., 1992).

An advantage, especially when primary fibroblasts are used in an *ex vivo* gene therapy approach, is the possibility to generate cells from the same individual receiving them as an autologous graft, which in turn reduces the need for immuno-suppression (Kawaja et al., 1991). Primary fibroblasts survive 8 to 10 weeks after implantation into the brain (Kawaja et al., 1991; Fisher et al., 1991), and fibroblasts transduced with viral vectors harbouring the TH-cDNA showed transgene expression and compensation of motor impairment in the rodent model of PD for at least 8 weeks (Fisher et al., 1991).

More recently, primary fibroblasts were double transduced with the gene of TH as well as that for GTP-cyclohydrolase 1 (GTP-CH1) (Bencsics et al., 1996). GTP-CH1 is needed for the biosynthesis of tetrahydrobiopterin (BH4), which in turn serves as an essential co-factor to TH for hydroxylation of tyrosine. In contrast to the other enzymes needed for the *de novo* BH4 synthesis, GTP-CH1 is not constitutively expressed in fibroblasts (Werner et al., 1990). Only cells containing both genes were able to produce L-dopa sufficiently *in vitro* and *in vivo*. Unfortunately, transgene expression was not detectable longer than 14 days *in vivo* and no difference from controls was observed in apomorphin induced rotation experiments (Bencsics et al., 1996).

These results rise also doubts about other reports (Fisher et al., 1991), where implantation of primary fibroblasts genetically modified to express the TH transgene alone resulted in biochemical and functional improvements in animal models of PD, because BH4 levels in denervated striatum are not sufficient to support TH activity (Levine et al., 1981). As mentioned above, it is possible that functional improvements in the rotation model, at least in part, are due to unspecific effects caused by limited brain injury through the transplantation procedure (Wang et al., 1994; Giulian and Lachman, 1985).

Genetically modified minced autologous and syngeneic muscle grafts also have been transplanted intracerebrally (Jiao et al., 1992a, 1992b). These grafts survived longer than 6 months after transplantation into the brain (Jiao et al., 1992a, 1992b) and transgene expression in vivo was reported to be detectable for at least 2 months (Jiao et al., 1992a). However, the paper by Jiao et al. (1993) reporting long-term amelioration of rotational behaviour in a PD rat model by plasmid-transfected primary muscle cells, expressing TH transgene was retracted three years later (Wolff, 1996).

In conclusion, the transplantation of genetically modified non-neural cells, capable to produce L-dopa has yet failed to be an alternative to the transplantation of foetal nigral tissue, at least at the present time.

Ex vivo gene therapy for neuroprotection

Neurotrophic factors including brain-derived neurotrophic factor (BDNF) (Hyman et al., 1991), neurotrophin-4/5 (NT4/5) (Hyman et al., 1994), ciliary neurotrophic factor (CNTF) (Hagg and Varon, 1993) and glial cell line-derived neurotrophic factor (GDNF) (Lin et al., 1993) support the survival of dopaminergic neurones, enhance their function and partially protect them against toxin-induced damage (Bilang-Bleuel et al., 1997; Choi-Lundberg et al., 1997). Since these factors do not pass the blood-brain barrier, it is crucial to deliver them intraventricularly or intracerebrally to target neurones within the brain (Björklund et al., 1997).

BDNF producing immortalised fibroblasts have been transplanted dorsal to the SN seven days prior to 1-methyl-4-phenylpyridinium (MPP$^+$) infusion in rats. Functional and histological evaluation showed a marked amelioration of amphetamine induced rotational asymmetry and increased survival of dopaminergic neurones compared to controls (Frim et al., 1994). Levivier et al. (1995) transplanted BDNF producing primary fibroblasts into the striatum 2 weeks before 6-OHDA lesioning with similar results. In addition, primary astrocytes genetically modified to produce BDNF (Yoshimoto et al., 1995) and NGF (Cunningham et al., 1991) have also proved to be protective in the rat model of PD.

It is important to keep in mind that it is unclear whether current animal models using the neurotoxin 6-OHDA, which causes a more or less acute degeneration of nigrostriatal dopaminergic neurones (Kelly et al., 1975; Sauer and Oertel, 1994) are comparable with the chronic degenerative processes in humans, which are caused by toxins alone only in a small minority of cases (Duvoisin, 1992). In addition, it is unknown, whether a neuroprotective therapy starting at the time of onset of PD in human beings is sufficient to slow down degeneration of dopaminergic neurones or to maintain them, since the DA content in the striatum is already reduced by about 70–80% and at least 50% of the dopaminergic neurones in the SN are already lost (Fearnley and Lees, 1991).

Another application for a neuroprotective approach is the co-transplantation of cells producing neurotrophic factors with embryonic ventral mesencephalon grafts. Two strategies have been evaluated successfully: (1) Transplantation of a mixture of basic fibroblast growth factor (bFGF) producing fibroblasts with foetal dopaminergic neurones (Takayama et al., 1995) and (2) co-transplantation of GDNF producing polymer encapsulated baby hamster kidney cells with solid ventral mesencephalon culture grafts (Sautter et al., 1998). Secreted growth factors enhanced the survival and the fibre out-growth of grafted dopaminergic neurones with a significant reduction of rotation asymmetry in the rodent model of PD in both studies (Takayama et al., 1995; Sautter et al., 1998).

Ex vivo gene therapy using CNS-derived progenitor cells

A relatively new strategy in the field of *ex vivo* gene therapy is the propagation and manipulation of neural precursor cells (Gage, 1998). Several studies showed that precursor cells share important features with stem cells, including pluripotency and selfrenewal capacity (Reynolds and Weiss, 1996; Weiss et al., 1996). Precursor cells could therefore serve as a source of cells generated to express an appropriate neuronal phenotype with the aim to substitute degenerated neurones i.e. dopaminergic neurones in PD.

In general, propagation of primary neuronal precursor cells is possible in two different ways: (1) stimulation with mitogens including epidermal growth factor (EGF) (Mytilineou et al., 1992; Svendsen et al., 1995; Hulspas et al., 1997; Shetty and Turner, 1998), bFGF (Buc-Caron, 1995; Gage et al., 1995; Mayer et al., 1993; Sabate et al., 1995) or with both (Weiss et al., 1996), and (2) transformation of the precursor cells by conditional oncogenes (Frederiksen et al., 1988; Snyder et al., 1992; Onifer et al., 1993; Anton et al., 1994).

Growth factor expanded neural progenitors have been obtained from embryonic (Mayer et al., 1993; Mytilineou et al., 1992; Svendsen et al., 1995) as well as from adult (Gage et al., 1995; Reynolds and Weiss, 1992; Richards et al., 1992) brain regions and have been shown to survive intracerebral transplantation into the rat brain (Gage et al., 1995; Svendsen et al., 1996, 1997).

Human neural progenitor cells have been transplanted into the striatum of unilaterally 6-OHDA lesioned adult rats (Svendsen et al., 1997). Transplanted progenitor cells were seen to differentiate into neurones and glia cells and survived for up to 20 weeks in the rat brain, although the majority of transplants had reduced in size to a thin strip of cells at this time point. This was possibly due to migration away from the core of the transplant (Svendsen et al., 1997). In contrast, other studies revealed only a 3- and 4-week survival time after transplantation respectively (Sabate et al., 1995; Svendsen et al., 1996). Interestingly, in only 2 out of 40 animals a small number of TH-positive neurones could be detected and in these 2 animals only, a reduction of amphetamine induced rotation has been observed after 20 weeks post grafting (Svendsen et al., 1997).

Temperature-sensitive immortalised neural progenitor cells have been genetically modified with TH cDNA resulting in an increase of L-DOPA production *in vitro* and in a significant decrease in apomorphine-induced rotation asymmetry (Anton et al., 1994). In addition, neural progenitors propagated with bFGF and transduced with a viral vector encoding the marker gene beta-galactosidase showed transgene expression in only 4 of 13 rats after transplantation (Sabate et al., 1995).

Transplantation of genetically modified neuronal precursor cells may be the most promising novel approach for *ex vivo* gene therapy: In principal, it is possible to propagate, genetically modify and to screen neural precursor cells after explantation *in vitro* (Frederiksen et al., 1988; Mehler et al., 1993). To some extent, various neuronal phenotypes can be established by inducing effects of neurotrophic factors either *in vitro* (Shetty and Turner, 1998) or *in vivo*, when precursor cells are transplanted to specific sites (Gage et al., 1995; Snyder et al., 1992; Svendsen et al., 1997). Also, they are capable of forming proper connections to host neurones (Gage et al., 1995; Snyder et al., 1992). Finally, with improved propagation of human neural progenitors in combination with appropriate genetic modifications, neural tissue from a single foetus may be sufficient to graft several patients, also providing the possibility to screen them for any contaminating pathogens prior to transplantation (Raymon et al., 1997).

In conclusion, first steps towards an application of neuronal precursor cells in models of PD have been made but many problems are still to be solved including induction of a distinct neuronal phenotype and long-term survival after transplantation.

Xenotransplantation for *ex vivo* gene therapy

More recently, efforts have been made to use pig and bovine tissue for xenotransplantation (Deacon et al., 1997; Zawada et al., 1998). A major problem of cross species transplantation is the "hyperacute rejection" of donor tissue (Dorling et al., 1997; Platt, 1998). This immune phenomenon is accompanied by a complement dependent destruction of vascular endothelia of the graft within minutes after transplantation, triggered by naturally occurring antibodies against galactose-alpha(1–3)-galactose (Weiss, 1998; Platt, 1998). To overcome this problem, tissue from transgenic pigs has been generated, that expresses human complement regulatory proteins, including CD46, CD55 and CD59 (Heckl et al., 1995; Platt, 1994, 1998). Knockout pigs for galactose-alpha(1–3)-galactose are not yet available, but knockout mice showed a reduced binding and immuno-reactivity to human serum (Tearle et al., 1996).

There are several concerns in the research community about this new strategy: Elimination of some animal viruses in humans is triggered by binding of antibodies against galactose-alpha(1–3)-galactose residues on the viral envelope (Rother et al., 1995; Takeuchi et al., 1996). In consequence, modification of pig tissue that protects the graft against hyperacute rejection, may lead to a resistance of enveloped viruses against complement mediated lysis (Patience et al., 1997). Another point to consider is that human complement regulatory proteins serve as receptors for some human viruses (Dörig et al., 1993; Ward et al., 1994).

Despite these concerns, Schumacher et al. transplanted embryonic ventral mesencephalon of normal, not genetically manipulated pigs into the striatum of 12 patients with PD (Schumacher, manuscript in preparation). Histopathological findings of one patient, who died of a pulmonary embolism 7 months after pig cell transplantation have been published (Deacon et al., 1997). In this patient immuno-suppression with cyclosporine prior and after transplantation was sufficient to achieve long-term graft survival over seven months and fibre out-growth of the transplanted pig neurones into host tissue (Deacon et al., 1997).

These and other results from animal studies suggest, that immunological reactions against intracerebrally transplanted neural xenografts are relatively mild compared to that seen in extracerebral xenotransplantations (Sachs and Bach, 1990; Steele and Auchincloss, 1995). This raises the question if there is a need to use genetically manipulated pig tissue for intracerebral transplantation at all.

In vivo gene therapy in Parkinson's disease

In vivo gene therapy in PD consists of direct application of vector DNA into the brain. The most effective way to achieve gene delivery to CNS-tissue is to

Table 1. Comparison of cell systems used for *ex vivo* gene therapy approaches in PD

	Advantages	Disadvantages	Remarks
Cell lines	indefinite supply easy characterisation and standardisation easy to modify	no long-term survival *in vivo* tumuor formation rejection	obsolete for intra-cerebral grafting (exception: "en-capsulation approach")
Primary cells (non-neuronal cells)	autologous trans-plantation possible long-term survival *in vivo*	limited supply no projection to adjacent host tissue limited trangene production *in vivo*	
Progenitor cells	autologous trans-plantation possible projection to adjacent host tissue	difficulties to induce dopaminergic pheno-type embryonic tissue as source of cells	probably the most promising new approach
Pig cells	good supply projection to adjacent host tissue	ethical considerations transspecies infections	very controversial in the research community trials in humans already exist

use virus-based vector systems. Several types of viral vectors have been constructed to express transgenes including the TH gene (During et al., 1994; Horellou et al., 1994; Kaplitt et al., 1994) and genes coding for neurotrophic factors, particularly GDNF (Bilang-Bleuel et al., 1997; Choi-Lundberg et al., 1997). The vector systems most often used for delivering foreign genes into neural cells are based on herpes simplex virus 1 (HSV-1), adenovirus (Ad), adeno-associated virus (AAV), and more recently retroviruses.

HSV-1

HSV-1 wild-type virus is a 152 kb dsDNA neurotropic human virus, that either enters a lytic cycle or alternatively establishes a latent state in neurones after infection with life-long persistence of the extrachromosomal viral genome as a circular episomal molecule (Speck and Simmons, 1991). Replication-defective deletion mutants of HSV-1 with the ability to maintain this latent state but lacking the potential to cause active inflammation have been constructed (Fink et al., 1992; Meignier et al., 1988). These virus vectors are capable to carry transgenes up to 30 kb (Glorioso et al., 1994).

Alternatively, amplicons consisting of an HSV origin of replication, HSV packaging signals, and a bacterial origin of replication, have been constructed (Spaete and Frenkel, 1982). Unfortunately, a defective HSV-1 virus is needed to package these amplicons, resulting in the risk of contaminating viral stocks and the emergence of wild-type recombinants due to repeated passaging (Glorioso et al., 1994).

A major problem of replication defective HSV-1 vectors is its residual cytotoxic effect after application into the brain (Karpati et al., 1996). To overcome this problem certain deletion mutants were created that lack expression of HSV gene products, especially immediate early antigens (IE), suspected to cause cytotoxicity (Glorioso et al., 1994). Long-term gene expression is also a point to consider when a HSV-1 based vector is used, since special promotor elements are needed to maintain transgene production in the latent state (Glorioso et al., 1994; Ramakrishnan et al., 1994).

Cultured striatal cells transduced with HSV-1 based vector constructs expressing TH cDNA (pHSVth) showed release of L-dopa and dopamine into the culture medium (Geller et al., 1995). In addition, amelioration of 6-OHDA induced rotation asymmetry in the rat model of PD for up to one year was observed after intrastriatal application of pHSVth (During et al., 1994).

AAV

Wild-type adeno-associated-virus (AAV) is a non-pathogenic single-stranded DNA parvovirus, that needs helper functions from other viruses for replication and reproduction (Berns and Bohenzky, 1987; Xiao et al., 1997). One striking feature of wild-type AAV is its unique ability as a DNA virus to integrate into a specific site of the hosts chromosome 19 (Samulski et al., 1989). In order to create AAV based viral vectors, more than 90% of the wild-type AAV genome have been deleted so that the recombinant AAV based (rAAV) vectors consist only of inverted terminal-repeat sequences, containing DNA replication sequences and packaging signals (Kaplitt et al., 1994; Samulski et al., 1989). These multiple deletions prevent the generation of wild-type virus (Samulski et al., 1989) and the expression of immunogenic viral gene products that could possibly trigger immune reactions of the host (Afione et al., 1996; Xiao et al., 1996).

Recombinant AAV vectors are able to transduce a very broad range of dividing and nondividing cells, including neurones and glia cells (Afione et al., 1996). According to the persistence of the vector DNA in the target cells there are results suggesting that rAAV integrates in chromosomal DNA as seen in wild type AAV (Xiao et al., 1996) but there are also studies indicating a persistence of the vector as episomal DNA (Afione et al., 1996).

The major disadvantage of this vector type is its limitation to package transgenes not larger than approximately 5 kb (Freese et al., 1996). Another problem is the need of adenovirus to generate rAAV, which may lead to contamination of vector preparations, but appropriate steps have been made to overcome this drawback by creating a plasmid, encoding the necessary Ad helper genes for rAAV replication (Ferrari et al., 1996).

In an *in vivo* gene therapy approach using rAAVth, a therapeutic effect was detected both in a rat model of Parkinson's disease (Kaplitt et al., 1994) as well as in monkeys with 1-methyl-4-phenyl-1,2,3,6-tetrahydropyridine (MPTP)-induced parkinsonism (During et al., 1998).

Adenoviruses

Recombinant adenovirus (rAd) vectors are most commonly derived from type 5 Ad so that the probability of integration of this DNA virus into chromosomal DNA is only very low (Horellou et al., 1997). To create a recombinant Ad vector it has proven essential to delete several genes of wild-type Ad virus to prevent a lytic infection (Graham et al., 1977, Anderson, 1998).

In contrast to rAAV vectors, recombinant Ad vectors are capable of carrying transgenes up to 35 kb in length (Anderson, 1998). As rAAV vectors, recombinant Ad vectors can transduce dividing and nondividing cells, including neurones and glia cells (Akli et al., 1993). No helper-virus is needed for the propagation of rAd vectors and very high vector concentrations are attainable (Karpati et al., 1996).

A possible limitation of the use of Ad based vectors is the immunological reaction after intracerebral application of high virus titers resulting in neural death and vascular inflammation (Akli et al., 1993; Le Gal La Salle et al., 1993). Nevertheless, intracerebral transgene expression mediated through recombinant Ad vectors was detected for months, probably because only low virus titers are needed to achieve sufficient gene expression and because the

CNS is in part an organ with less stringent immune surveillance (Le Gal La Salle et al., 1993).

Recombinant Ad vectors harbouring the TH gene have been used for intrastiatal injection in the rat model of PD resulting in a significant decrease of apomorphin induced turning behaviour one and two weeks after vector application (Horellou et al., 1994).

More recently replication-defective Ad vectors encoding GDNF have been used in the rat model of PD created by Sauer and Oertel in which 6-OHDA is injected into the striatum leading to a relatively slow retrograde degeneration of dopaminergic neurones (Sauer and Oertel, 1994). One week prior to intrastriatal 6-OHDA injection, Ad-GDNF was injected into the striatum (Bilang-Bleuel et al., 1997) or immediately dorsal to the substantia nigra (Choi-Lundberg et al., 1997). In both experimental settings, a significant increase in the survival of the dopaminergic neurones after toxin application compared to controls was detected and a highly significant reduction of amphetamine induced rotation asymmetry was observed in the first study whereas no behavioural data have been shown in the latter.

Retroviruses

More recently, several lentivirus-based retroviral vector systems are under investigation for an *in vivo* gene therapy approach. The lentivirus family is one subclass of retroviruses including the HIV virus. In contrast to other retroviral vectors such as murine leukemia virus, HIV-1 based vectors are able to transduce nondividing cells, including adult neurones (Naldini et al., 1996b). Integration into the host genome is established by expression of karyophilic determinants which interact with the nuclear import machinery that in turn enables the entry of HIV preintegration complex into the nucleus of target cells (Bukrinsky et al., 1993).

Attenuated replication-incompetent HIV-1 vectors have been constructed by deleting up to 5 out of 6 auxiliary genes constitutively present in wild-type HIV-1 virus (Kim et al., 1998; Zufferey et al., 1997). The full function compared to wild-type HIV-1 could be preserved, whereas transduction efficiency of these lentiviral vector constructs declined by 50% (Zufferey et al., 1997). Deletion of accessory genes is crucial, since expression of these genes are under suspicion to cause Kaposi's sarcoma, cell cycle arrest and apoptosis (Kim et al., 1998). In addition, to obtain higher vector titers and to broaden the host range, the env gene of wild-type HIV-1 can be exchanged with a gene coding for surface glycoproteins of other viruses, especially vesicular stomatitis virus (VSV), leading to VSV G-pseudotyped lentiviral vectors (Naldini et al., 1996a, 1996b).

Replication-defective pseudotyped lentivirus vectors, harbouring the beta-gal marker gene were injected into rat striatum and hippocampus (Naldini et al., 1996b). After a 30-day period beta-gal positive neurones and glia cells could be detected (Naldini et al., 1996b), suggesting that *in vivo* gene

Table 2. Comparison of vector systems for *in vivo* gene therapy approaches in PD

	Advantages	Disadvantages	Remarks
HSV-1	maximum transgene size 35 kb persists in latent state	cytopathogenicity	
AAV	integration into host genome (site specific?) long-term transgene expression *in vivo*	difficult preparation contamination with Adenovirus (due to preparation) carries transgenes not larger than 5 kd	
Adenovirus	maximum transgene size 35 kb easy to generate	causes inflammation and is cytotoxic	vector system with most severe immunological problems
Retrovirus (Lentivirus)	integration into host genome long-term transgene expression *in vivo*	risk of viral rearrangements mutagenesis due to integration	newest vector system with great potential

delivery by HIV-based vector systems is a possible alternative to other viral vector systems described above.

Despite all advantages of HIV-1 based vectors, there are still concerns about viral rearrangements leading to wild-type HIV-1 virus generation, even though this would be a very unlikely event.

Conclusion

At the present time, transplantation of foetal nigral tissue is still the "gold standard" in transplantation treatment of PD in humans, although the outcome of the studies are rather inconsistent (Freed et al., 1992; Wenning et al., 1997). The emergence of ethical and religious concerns, particularly in the U.S. was one major reason to search for alternative tissue sources and therefore to initiate *ex vivo* gene therapy experiments under way using tissue of non-foetal origin. After 10 years of research there are still no clinical trials under way using genetically modified tissue or cells for transplantation in patients with PD, and several major problems still remain to be solved: (1) Safety of vector-systems, (2) survival and "fate" of transplanted tissue, (3) long-term gene expression within the transplant, and (4) immune response of the host. As in *ex vivo* gene therapy, no *in vivo* gene therapy approach for PD has reached clinical relevance until now.

New directions in *ex vivo* gene therapy by using neuronal precursor cells still have to prove their effectiveness and many problems, such as long-term survival after intracerebral transplantation have to be solved. Transplantation of xenogenic neural cells may be another possible solution, but again, further studies are needed especially because of the hazard to transmit animal pathogens to humans.

Acknowledgement

This work is supported by the BMBF grant 01KV9511/7.

References

Afione SA, Conrad CK, Kearns WG, Chunduru S, Adams R, Reynolds TC, Guggino WB, Cutting GR, Carter BJ, Flotte TR (1996) In vivo model of adeno-associated virus vector persistence and rescue. J Virol 70: 3235–3241

Akli S, Caillaud C, Vigne E, Startford-Perricaudet LD, Poenaru L, Perricaudet M, Kahn A, Peschanski MR (1993) Transfer of foreign gene into the brain using adenovirus vectors. Nat Gen 3: 224–228

Anderson WF (1998) Human gene therapy. Nature 392: S25–S30

Anton R, Kordower JH, Maidment NT, Manaster JS, Kane DJ, Rabizadeh S, Schueller SB, Yang J, Edwards RH, Markham CH, Bredesen DE (1994) Neural-targeted gene therapy for rodent and primate hemiparkinsonism. Exp Neurol 127: 207–218

Bencsics C, Wachtel RS, Milstien S, Kang UJ (1996) Double transduction with GTP cyclohydrolase 1 and tyrosine hydroxylase is necessary for spontaneous synthesis of L-dopa by primary fibroblasts. J Neurosci 16: 4449–4456

Berns KI, Bohenzky RA (1987) Adeno-associated viruses: an update. Adv Virus Res 32:243–306

Bilang-Bleuel A, Revah F, Colin P, Locquet I, Robert JJ, Mallet J, Horellou P (1997) Intrastriatal injection of an adenoviral vector expressing glial-cell-line-derived neurotrophic factor prevents dopaminergic neuron degeneration and behavioural impairment in a rat model of Parkinson disease. Proc Natl Acad Sci USA 94: 8818–8823

Björklund A, Rosenblad C, Winkler C, Kirik D (1997) Studies on neuroprotective and regenerative effects of GDNF in a partial lesion model of Parkinson's disease. Neurobiol Dis 4: 186–200

Buc-Caron MH (1995) Neuroepithelial progenitor cells explanted from human fetal brain proliferate and differentiate in vitro. Neurobiol Dis 2: 37–47

Bukrinsky MI, Haggerty S, Dempsey MP, Sharova N, Adzhubei A, Spitz L, Lewis P, Goldfarb D, Emerman M, Stevenson M (1993) A nuclear localization signal within HIV-1 matrix protein that governs infection of non-dividing cells. Nature 365: 666–669

Choi-Lundberg D, Lin Q, Chang Y, Chiang YL, Hay CM, Mohajeri H, Davidson BL, Bohn MC (1997) Dopaminergic neurons protected from degeneration by GDNF gene therapy. Science 275: 838–841

Cunningham LA, Hansen JT, Short MP, Bohn MC (1991) The use of genetically altered astrocytes to provide nerve growth factor to adrenal chromaffin cells grafted into the striatum. Brain Res 561: 192–202

Deacon T, Schumacher J, Dinsmore J, Thomas C, Palmer P, Isacson O (1997) Histological evidence of fetal pig neural cell survival after transplantation into a patient with Parkinson's disease. Nat Med 3: 350–353

Dorling A, Riesbeck K, Warrens A, Lechler R (1997) Clinical xenotransplantation of solid organs. Lancet 349: 867–871

Dörig RE, Marcil A, Chopra A, Richardson CD (1993) The human CD46 molecule is a receptor for measles virus (Edmonston Strain). Cell 75: 295–305

During MJ, Naegele JR, O'Malley KL, Geller AI (1994) Long-term behavioral recovery in parkinsonian rats by an HSV vector expressing tyrosine hydroxylase. Science 266: 1399–1403

During MJ, Samulski RJ, Elsworth JD, Kaplitt MG, Leone P, Xiao X, Li J, Freese A (1998) In vivo expression of therapeutic human genes for dopamine production in the caudates of MPTP-treated monkeys using an AAV vector. Gene Ther 5: 820–827

Duvoisin R (1992) Overview of Parkinson's disease. Ann N Y Acad Sci 648: 187–193

Fearnley JM, Lees AJ (1991) Aging and Parkinson's disease: Substantia nigra regional specificity. Brain 114: 2283–2301

Ferrari FK, Samulski T, Shenk T, Samulski RJ (1996) Second-strand synthesis is a rate-limiting step for efficient transduction by recombinant adeno-associated virus vectors. J Virol 70: 3227–3234

Fink DJ, Sternberg PC, Weber M, Mata M, Goins WF, Glorioso JC (1992) In vivo expression of beta-galactosidase in hippocampal neurons by HSV-mediated gene transfer. Hum Gene Ther 3: 11–19

Fisher L, Jinnah H, Kale L, Higgins G, Gage F (1991) Survival and function of intrastriatally grafted primary fibroblasts genetically modified to produce L-DOPA. Neuron 6: 371–380

Forno LS (1996) Neuropathology of Parkinson's disease. J Neuropath Exp Neurol 55: 259–272

Frederiksen K, Jat PS, Valtz N, Levy D, McKay R (1988) Immortalization of precursor cells from the mammalian CNS. Neuron 1: 439–448

Freed CR, Breeze RE, Rosenberg NL, Schneck SA, Kriek E, Qi JX, Lone T, Zhang YB, Snyder JA, Wells TH, et al (1992) Survival of implanted fetal dopamine cells and neurological improvement 12 to 46 months after transplantation for Parkinson's disease. N Engl J Med 327: 1549–1555

Freese A, Stern M, Kaplitt MG, O'Connor WM, Abbey MV, O'Connor MJ, During MJ (1996) Prospects for gene therapy in Parkinson's disease. Mov Disord 11: 469–488

Frim DM, Uhler TA, Galpern WR, Beal MF, Breakfield XO, Isacson O (1994) Implanted fibroblasts genetically engineered to produce brain-derived neurotrophic factor prevent MPTP toxicity to dopaminergic neurons in the rat. Proc Natl Acad Sci USA 91: 5104–5108

Gage F (1998) Cell therapy. Nature 392: S18–S24

Gage F, Coates PW, Palmer TD, Kuhn HG, Fisher L, Suhonen JO, Peterson DA, Suhr ST, Ray J (1995) Survival and differentation of adult neuronal progenitor cells transplanted to the adult brain. Proc Natl Acad Sci USA 92: 11879–11883

Geller AI, Freese A (1990) Infection of cultured central nervous system neurons with a defective herpes simplex virus 1 vector results in stable expression of E. coli beta-galactosidase. Proc Natl Acad Sci USA 87: 1149–1153

Geller AI, Freese A, During MJ, O'Malley KL (1995) A HSV-1 vector expressing tyrosine hydroxylase causes production and release of L-dopa from cultured rat striatal cells. J Neurochem 64: 487–496

Giulian D, Lachman LB (1985) Interleukin-1 stimulation of astroglial proliferation after brain injury. Science 228: 497–499

Glorioso JC, Goins WF, Meaney CA, Fink DJ, DeLuca NA (1994) Gene transfer to brain using herpes simplex virus vectors. Ann Neurol 35: S28–S34

Graham FL, Smiley J, Russell WC, Nairn R (1977) Characteristics of a human cell line transformed by DNA from human adenovirus 5. J Gen Virol 36: 59–72

Hagg T, Varon S (1993) Ciliary neurotrophic factor prevents degeneration of adult rat substantia nigra dopaminergic neurons in vivo. Proc Natl Acad Sci USA 90: 6315–6319

Heckl-Ostreicher B, Binder R, Kirschfink M (1995) Functional activity of the membran-associated complement inhibitor CD59 in a pig to human in vitro model for hyper-acute xenograft rejection. Clin Exp Immunol 102: 589–595

Hefti F, Hartikka J, Schlumpf M (1985) Implantation of PC12 cells into the corpus striatum of rats with lesions of the dopaminergic nigrostriatal neurons. Brain Res 348: 283–288

Horellou P, Guibert B (1989) Retroviral transfer of a human tyrosine hydroxylase cDNA in various cell lines: regulated release of dopamine in mouse anterior pituitary AtT-20 cells. Proc Natl Acad Sci USA 86: 7233–7237

Horellou P, Brundin P, Kalen P, Mallet J, Björklund A (1990a) In vivo release of DOPA and dopamine from genetically engineered cells grafted to the denervated rat striatum. Neuron 5: 393–402

Horellou P, Marlier L, Privat A, Mallet J (1990b) Behavioural effect of engineered cells that synthesize L-DOPA or dopamine after grafting into the rat neostriatum. Eur J Neurosci 2: 116–119

Horellou P, Marlier L, Privat A, Mallet J (1990c) Exogenous expression of L-dopa and dopamine in various cell lines following transfer of rat and human tyrosine hydroxylase cDNA: Grafting in an animal model of Parkinson's disease. Prog Brain Res 82: 23–32

Horellou P, Vigne E, Castel MN, Barneoud P, Colin P, Perricaudet M, Delaere P, Mallet J (1994) Direct intracerebral gene transfer of an adenoviral vector expressing tyrosine hydroxylase in a rat model of Parkinson's disease. Neuroreport 6: 49–53

Horellou P, Sabate O, Buc-Caron MH, Mallet J (1997) Adenovirus-mediated gene transfer to the central nervous system for Parkinson's disease. Exp Neurol 144: 131–138

Hulspas R, Tiarks C, Reilly J, Hsieh CC, Recht L, Quesenberry PJ (1997) In vitro cell density-dependent clonal growth of EGF-responsive murine neural progenitor cells under serum-free conditions. Exp Neurol 148: 147–156

Hyman C, Hofer M, Barde YA, Juhasz M, Yancopoulos GD, Squinto SP, Lindsay RM (1991) BDNF is an neurotrophic factor for dopaminergic neurons of the substantia nigra. Nature 350: 230–232

Hyman C, Juhasz C, Jackson C, Wright P, Ip NY, Lindsay RM (1994) Overlapping and distinct actions of the neurotrophins BDNF, NT-3 and NT-4/5 on cultured dopaminergic and GABAergic neurons of the ventral mesencephalon. J Neurosci 14: 335–347

Jiao S, Schultz E, Wolff J (1992a) Intracerebral transplants of primary muscle cells: A potential "platform" for transgene expression in the brain. Brain Res 575: 143–147

Jiao S, Wolff J (1992b) Long-term survival of autologous muscle grafts in rat brain. Neurosci Lett 137: 207–210

Jiao S, Gurevich V, Wolff J (1993) Long-term correction of rat model of Parkinson's disease by gene therapy. Nature 362: 450–453

Kaplitt MG, Leone P, Samulski RJ, Xiao X, Pfaff DW, O'Malley KL, During MJ (1994) Long-term gene expression and phenotypic correction using adeno-associated virus vectors in the mammalian brain. Nat Gen 8: 148–154

Karpati G, Lochmüller H, Nalbantoglu J, Durham H (1996) The principles of gene therapy for the nervous system. Trends Neurosci 19: 49–54

Kawaja M, Gage F (1992) Morphological and neurochemical features of cultured primary skin fibroblasts of Fischer 344 rats following striatal implantation. J Comp Neurol 317: 102–116

Kawaja M, Fagan AM, Firestein BL, Gage F (1991) Intracerebral grafting of cultured autologous skin fibroblasts into rat striatum: An assessment of graft size and ultrastructure. J Comp Neurol 307: 695–706

Kelly PH, Seviour PW, Iversen SD (1975) Amphetamine and apomorphine responses in the rat following 6-OHDA lesions of the nucleus accumbens septi and corpus striatum. Brain Res 94: 507–522

Kim VN, Mitrophanous K, Kingsman SM, Kingsman AJ (1998) Minimal requirement for a lentivirus vector based on human immundeficiency virus type 1. J Virol 72: 811–816

Kordower JH, Goetz CG, Freeman TB, Olanow CW (1997) Dopaminergic transplants in patients with Parkinson's disease: Neuroanatomical correlates of clinical recovery. Exp Neurol 144: 41–46

Lange K, Youdim M, Riederer P (1992) Neurotoxicity and neuroprotection in Parkinson's disease. J Neural Transm 38: 27–44

Le Gal La Salle G, Robert JJ, Berrard S, Ridoux V, Startford-Perricaudet LD, Perricaudet M, Mallet J (1993) An adenovirus vector for gene transfer into neurons and glia in the brain. Science 259: 988–990

Levine RA, Miller LP, Lovenberg W (1981) Tetrahydrobiopterin in striatum: localization in dopamine nerve terminals and role in catecholamine synthesis. Science 213: 349–350

Levivier M, Przedborski S, Bencsics C, Kang UJ (1995) Intrastriatal implantation of fibroblasts genetically engineered to produce brain-derived neurotrophic factor prevents degeneration of dopaminergic neurons in rat model of Parkinson's disease. J Neurosci 15: 7810–7820

Lin LF, Doherty DH, Lile JD, Bektesh S, Collins F (1993) GDNF: A glial cell line-derived neurotrophic factor for midbrain dopaminergic neurons. Science 260: 1130–1132

Lundberg C, Horellou P, Mallet J, Björklund A (1996) Generation of dopa-producing astrocytes by retroviral transduction of the human tyrosine hydroxylase gene: *in vitro* characterization and in vivo effects in the rat Parkinson model. Exp Neurol 139: 39–53

Mayer E, Dunnett SB, Fawcett JW (1993) Mitogenic effect of basic fibroblast growth factor on embryonic ventral mesencephalic dopaminergic neuron precursors. Brain Res Dev Brain Res 72: 253–258

Mehler MF, Rozental R, Dougherty M, Spray DC, Kessler JA (1993) Cytokine regulation of neural differentiation of hippocampal progenitor cells. Science 362: 62–65

Meignier B, Longnecker R, Mavromara-Nazos P, Sears A, Roizman B (1988) Virulence of and establishment of latency by genetically engineered deletion mutants of herpes simplex virus type 1. Virology 162: 251–254

Mytilineou C, Park TH, Shen J (1992) Epidermal growth factor-induced survival and proliferation of neuronal precursor cells from embryonic rat mesencephalon. Neurosci Lett 135: 62–66

Naldini L, Blomer U, Gage F, Trono D, Verma IM (1996a) Efficient transfer, integration, and sustained long-term expression of the transgene in adult rat brains injected with a lentiviral vector. Proc Natl Acad Sci USA 93: 11382–11388

Naldini L, Blömer U, Gallay P, Ory D, Mulligan R, Gage F, Verma IM, Trono D (1996b) *In vivo* gene delivery and stable transduction of nondividing cells by lentiviral vector. Science 272: 263–267

Onifer S, Whittemore SR, Holets VR (1993) Variable morphological differentation of a rapid-derived neuronal cell line following transplantation into the adult rat CNS. Exp Neurol 122: 130–142

Partridge TA, Davies KE (1995) Myoblast-based gene therapies. Br Med Bull 51: 123–137

Patience C, Takeuchi Y, Weiss RA (1997) Infection of human cells by an endogenous retrovirus of pigs. Nat Med 3: 282–286

Platt JL (1994) A perspective on xenograft rejection and accommodation. Immunol Rev 141: 127–149

Platt JL (1998) New directions for organ transplantation. Nature 392: S11–S17

Ramakrishnan R, Fink DJ, Guihua J, Desai P, Glorioso JC, Levine M (1994) Competitive quantitative polymerase chain reaction (PCR) analysis of herpes simplex virus typ 1 DNA and LAT RNA in latently infected cells of the rat brain. J Virol 68: 1864–1870

Raymon HK, Thode S, Gage F (1997) Application of ex vivo gene therapy in the treatment of Parkinson's disease. Exp Neurol 144: 82–91

Reynolds BA, Weiss S (1992) Generation of neurons and astrocytes from isolated cells of the adult mammalian central nervous system. Science 255: 1707–1710

Reynolds BA, Weiss RA (1996) Clonal and population analyses demonstrate that an EGF-responsive mammalian embryonic CNS precursor is a stem cell. Dev Biol 175: 1–13

146 M. Bauer et al.

Richards LJ, Kilpatrick TJ, Bartlett PF (1992) *De novo* generation of neuronal cells from the adult mouse brain. Proc Natl Acad Sci USA 89: 8591–8595

Rother RP, Fodor WL, Springhorn JP, Birks CW, Setter E, Sandrin MS, Squinto SP, Rollins SA (1995) A novel mechanism of retrovirus inactivation in human serum mediated by anti-alpha galactosyl natural antibody. J Exp Med 182: 1345–1355

Sabate O, Horellou P, Vigne E, Colin P, Perricaudet M, Buc-Caron MH, Mallet J (1995) Transplantation to the rat brain of human neural progenitors that were genetically modified using adenoviruses. Nat Gen 9: 256–260

Sachs DH, Bach FH (1990) Immunology of xenograft rejection. Hum Immunol 28: 245–251

Samulski JS, Chang LS, Shenk T (1989) Helper-free stocks of adeno-associated viruses: normal integration does not require viral gene expression. J Virol 63: 3822–3828

Sauer H, Oertel WH (1994) Progressive degeneration of nigrostriatal dopamine neurons following intrastriatal terminal lesions with 6-hydroxydopamine: a combined retrograde tracing and immunocytochemical study in the rat. Neuroscience 59: 401–415

Sautter J, Tseng JL, Braguglia D, Aebischer P, Spenger C, Seiler RW, Widmer HR, Zurn AD (1998) Implants of polymer-encapsulated genetically modified cells releasing glial cell line-derived neurotrophic factor improve survival, growth and function of dopaminergic grafts. Exp Neurol 149: 230–236

Shetty AK, Turner DA (1998) In vitro survival and differentation of neurons derived from epidermal growth factor-responsive postnatal hippocampal stem cells: inducing effects of brain-derived neurotrophic factor. J Neurobiol 35: 395–425

Snyder EY, Deitcher DL, Walsh C, Arnold-Aldea S, Hartwieg EA, Cepko CL (1992) Multipotent neural cell lines can engraft and participate in development of mouse cerebellum. Cell 68: 33–51

Spaete R, Frenkel N (1982) The herpes simplex virus amplicon: A new eucaryotic defective-virus cloning amplifying vector. Cell 30: 295–304

Speck PG, Simmons A (1991) Divergent molecular pathways of productive and latent infection with a virulent strain of herpes simplex virus type 1. J Virol 65: 4004–4005

Steele DJ, Auchincloss HJ (1995) Xenotransplantation. Annu Rev Med 46: 345–360

Svendsen CN, Fawcett JW, Bentlage C, Dunnett SB (1995) Increased survival of rat EGF-generated CNS precursor cells using B27 supplemented medium. Exp Brain Res 102: 407–414

Svendsen CN, Clarke DJ, Rosser AE, Dunnett SB (1996) Survival and differentiation of rat and human epidermal growth factor-responsive precursor cells following grafting into the lesioned adult central nervous system. Exp Neurol 137: 376–388

Svendsen CN, Caldwell MA, Shen J, ter Borg MG, Rosser AE, Tyers P, Karmiol S, Dunnett SB (1997) Long-term survival of human central nervous system progenitor cells transplanted into rat model of Parkinson's disease. Exp Neurol 148: 135–146

Takayama H, Ray J, Raymon HK, Baird A, Hogg J, Fisher L, Gage F (1995) Basic fibroblast growth factor increases dopaminergic graft survival and function in a rat model of Parkinson's disease. Nat Med 1: 53–58

Takeuchi Y, Porter CD, Strahan KM, Preece AF, Gustafsson K, Cosset FL, Weiss RA, Collins MKL (1996) Sensitization of cells and retroviruses to human serum by (alpha1–3) galactosyltransferase. Nature 379: 85–88

Tearle RG, Tange MJ, Zannettino ZL, Katerelos M, Shinkel TA, Van-Denderen BJ, Lonie AJ, Lyons I, Nottle MB, Cox MB, Becker MB, et al (1996) The alpha-1,3-galactosyltransferase knockout mouse: implications for xenotransplatation. Transplantation 61: 13–19

Wang J, Bankiewics KS, Plunkett RJ, Oldfield EH (1994) Intrastriatal implantation of interleukin-1. J Neurosurg 80: 484–490

Ward T, Pipkin PA, Clarkson NA, Stone DM, Minor PD, Almond JW (1994) Decay-accelerating factor CD55 is identified as the receptor for echovirus 7 using CELICS, a rapid immuno-focal cloning method. EMBO 13: 5070–5074

Weiss RA (1998) Transgenic pigs and virus adaptation. Nature 391: 327–328

Weiss RA, Reynolds BA, Vescovi AL, Morshead C, Craig CG, van der Kooy D (1996) Is there a neural stem cell in the mammalian forebrain? Trends Neurosci 19: 387–393

Wenning GK, Odin P, Morrish P, Rehncrona S, Widner H, Brundin P, Rothwell JC, Brown R, et al (1997) Short-and long-term survival and function of unilateral intrastriatal dopaminergic grafts in Parkinson's disease. Ann Neurol 42: 95–107

Werner ER, Werner-Felmayer G, Fuchs D, Hausen A, Reibnegger D, Yim JJ, Pfleiderer W, Wachter H (1990) Tetrahydrobiopterin biosynthetic activities in human macrophages, fibroblasts, THP-1, and T24 cells. J Biol Chem 265: 3189–3192

Wolff J (1996) Retraction of Jiao S, Gurevich V, Wolff JA, in Nature 1993 Apr 1;362 (6419):450–453. Nature 380: 734

Wolff J, Fisher L, Xu L, Gage F (1989a) Grafting fibroblasts genetically modified to produce L-dopa in a rat model of Parkinson disease. Proc Natl Acad Sci USA 86: 9011–9014

Wolff J, Fisher L, Xu L, Hyder A, Jinnah H (1989b) Grafting fibroblasts genetically modified to produce L-dopa in a rat model of Parkinson disease. Proc Natl Acad Sci USA 86: 9011–9014

Xiao X, Li J, Samulski JS (1996) Long-term and efficient in vivo gene transfer into muscle tissue of immunocompetent mice with an rAAV vector. J Virol 70: 8098–8108

Xiao X, Li J, McCown TJ, Samulski JS (1997) Gene transfer by adeno-associated virus vectors into the central nervous system. Exp Neurol 144: 113–124

Yoshimoto Y, Lin Q, Collier TJ, Frim DM, Breakfield XO, Bohn MC (1995) Astrocyts retrovirally transduced with BDNF elicit behavioural improvement in a rat model of Parkinson's disease. Brain Res 691: 25–36

Zawada WM, Cibelli JB, Choi PK, Clarkson ED, Golueke PJ, Witta SE, Bell KP, Kane J, Ponce De Leon FA, Jerry DJ, Robl JM, Freed CR, Stice SL (1998) Somatic cell cloned transgenic bovine neurons for transplantation in parkinsonian rats. Nat Med 4: 569–574

Zhu SM, Kujirai K, Dollison A, Angulo J, Fahn S, Cadet JL (1992) Implantation of genetically modified mesencephalic fetal cells into rat striatum. Brain Res Bull 29: 81–93

Zufferey R, Nagy D, Mandel RJ, Naldini L, Trono D (1997) Multiply attenuated lentiviral vector achieves efficient gene delivery in vivo. Nat Biotech 15: 871–875

Authors' address: PD Dr. med. T. Gasser, Neurologische Klinik, Klinikum Grosshadern, LMU München, Marchioninistrasse 15, D-81377 München, Federal Republic of Germany

SpringerNeurology

P. Riederer, D. B. Calne, R. Horowski, Y. Mizuno, W. Poewe, M. B. H. Youdim (eds.)

Advances in Research on Neurodegeneration

Volume 5

1997. VIII, 215 pages. 45 figures.
Hardcover DM 198,–, öS 1386,–
(recommended retail price). ISBN 3-211-82933-4
Special edition of "Journal of Neural Transmission, Suppl. 50, 1997"

Contents

Brain imaging revisited:
Magnetic Resonance: A multimodal approach to the brain? • Measurement of the dopaminergic degeneration in Parkinson's disease with [^{123}I]beta-CIT and SPECT • rCBF SPECT in Parkinson's disease patients with mental dysfunction • IBZM- and beta-CIT-SPECT of the dopaminergic system in parkinsonism • Pathophysiology of movement disorders studied using PET • Contributions of Positron Emission Tomography to elucidating the pathogenesis of Idiopathic Parkinsonism and Dopa Responsive Dystonia

Endogenous and exogenous neurotoxins:
Mechanism of 6-Hydroxydopamine neurotoxicity • Induction of mitosis-related genes during dopamine-triggered apoptosis in sympathetic neurons • Neuronal vulnerability in Parkinson's disease • N-Methyl-(*R*)salsolinol as a dopaminergic neurotoxin: From an animal model to an early marker of Parkinson's disease • The halogenated tetrahydro-beta-carboline "TaClo": A progressively-acting neurotoxin

Programmed cell death, apoptosis, necrosis and in between:
Developmental and genetic regulation of programmed neuronal death • Apoptosis in neurodegenerative disorders • Mechanisms of cell death in Alzheimer's disease • Assessment of neurotoxicity and "neuroprotection"

Immunoinflammatory mechanisms, infective diseases causing neurological disorders:
Update on management and genetics of multiple sclerosis • Pathogenesis of immune-mediated demyelination in the CNS • Basic mechanisms of brain inflammation • Cell death in prion disease

SpringerWienNewYork

Sachsenplatz 4–6, P.O.Box 89, A-1201 Wien, Fax +43-1-330 24 26, e-mail: books@springer.at, Internet: http://www.springer.at
New York, NY 10010, 175 Fifth Avenue • D-14197 Berlin, Heidelberger Platz 3 • Tokyo 113, 3–13, Hongo 3-chome, Bunkyo-ku

SpringerNeurology

H. J. Gertz, T. Arendt (eds.)

Alzheimer's Disease –
From Basic Research to Clinical Applications

1998. VIII, 315 pages. 52 figures.
Hardcover DM 198,–, öS 1386,–
ISBN 3-211-83113-4
Special edition of "Journal of Neural Transmission, Suppl. 54, 1998"

This volume brings together the reports of basic scientists and clinical investigators on Alzheimer's disease. The issue bridges the gap between laboratory work in basic science and the development of urgently needed therapeutic strategies. Areas presented are the molecular and cellular biology of the disease, pathogenetic mechanisms and potential therapeutic targets, genetics, risk factors, strategies of prevention and treatment as well as practical aspects of medical and social care for patients with Alzheimer's disease.

K. Jellinger, F. Fazekas, M. Windisch (eds.)

Ageing and Dementia

1998. VIII, 406 pages. 65 partly coloured figures.
Hardcover DM 248,–, öS 1736,–
ISBN 3-211-83115-0
Special edition of "Journal of Neural Transmission, Suppl. 53, 1998"

Ageing and dementia are closely related conditions. Increasing age of the general population causes increasing incidence of dementing disorders in later life, although cognitive impairment is not necessarily a consequence of advancing age. The book presents the papers of the International Symposium on Ageing and Dementia, October 17–19, 1997 in Graz, where internationally renowned experts in the field of ageing and dementia gave an overview of the current knowledge about the epidemiology, pathomorphology, clinical diagnosis and course of brain ageing processes and related dementing disorders, biochemical markers and imaging procedures for the diagnosis of Alzheimer's disease and current approaches to a successful treatment of dementia.

All prices are recommended retail prices

 SpringerWienNewYork

Sachsenplatz 4–6, P.O.Box 89, A-1201 Wien. Fax +43-1-330 24 26, e-mail: books@springer.at. Internet: http://www.springer.at
New York, NY 10010, 175 Fifth Avenue • D-14197 Berlin, Heidelberger Platz 3 • Tokyo 113, 3–13, Hongo 3-chome, Bunkyo-ku

SpringerNeurology

Y. Mizuno, M.B.H. Youdim, D.B. Calne, R. Horowski, W. Poewe, P. Riederer (eds.)

Advances in Research on Neurodegeneration

Volume 3 & 4

1997. VIII, 280 pages. 46 figures.
Hardcover DM 215,–, öS 1505,–
(recommended retail price). ISBN 3-211-82935-0
Special edition of "Journal of Neural Transmission, Suppl. 49, 1997"

Contents

Treatment strategies for neurodegenerative diseases based on trophic factors and cell transplantation techniques • Models of Alzheimer's disease: cellular and molecular aspects • The relationship of Alzheimer-type pathology to dementia in Parkinson's disease • Models to study the role of neurotrophic factors in neurodegeneration • Induction of experimental autoimmune encephalomyelitis by CD^4 T cells specific for an astrocyte protein, S100 beta • Immunological aspects of experimental allergic encephalomyelitis and multiple sclerosis and their application for new therapeutic strategies • Nigrostriatal neuronal death in Parkinson's disease – a passive or an active genetically-controlled process? • Animal model and in vitro studies of anti neurofilament antibodies mediated neurodegeneration in Alzheimer's disease • Cop 1 as a candidate drug for multiple sclerosis • Possible role of the cholinergic system and disease models • Loss of dopaminergic neurons in parkinsonism: possible role of reactive dopamine metabolites • Role of interferons in demyelinating disease • Role of interferons in demyelinating diseases • Mechanism of amyloid beta protein induced neuronal cell death: current concepts and future perspectives • From prion diseases to Alzheimer's disease • A phosphorylation cascade in the basal ganglia of the mammalian brain: regulation by the D-1 dopamine receptor. A mathematical model of known biochemical reactions • Molecular heterogeneity of neurotransporters: implications for neurodegeneration • Central nervous system cytokines and their relevance for neurotoxicity and apoptosis • Genetics of multiple sclerosis – how could disease-associated HLA-types contribute to pathogenesis? • Modulation of control mechanisms of dopamine-inducedapoptosis – a future approach to the treatment of Parkinson's disease? • GTP cyclohydrolase I gene, dystonia, juvenile parkinsonism, and Parkinson's disease • MRI as a method to reveal in-vivo pathology in MS • Familial amyotrophic lateral sclerosis • Chronic administration of a partial agonist at strychnine-insensitive glycine receptors: a novel experimental approach to the treatment of ischemias • Apoptosis in neurodegenerative disorders: potential for therapy by modifying gene transcription • Disinhibition-Dementia-Parkinsonism-Amyotrophy Complex (DDPAC) is a non-Alzheimer's frontotemporal dementia

SpringerWienNewYork

Sachsenplatz 4–6, P.O.Box 89, A-1201 Wien, Fax +43-1-330 24 26, e-mail: books@springer.at. Internet: http://www.springer.at
New York, NY 10010, 175 Fifth Avenue • D-14197 Berlin, Heidelberger Platz 3 • Tokyo 113, 3–13, Hongo 3-chome, Bunkyo-ku